Smart Innovation, Systems and Technologies

Volume 92

The Smart Innovation, Systems and Technologies book series encompasses the topics of knowledge, intelligence, innovation and sustainability. The aim of the series is to make available a platform for the publication of books on all aspects of single and multi-disciplinary research on these themes in order to make the latest results available in a readily-accessible form. Volumes on interdisciplinary research combining two or more of these areas is particularly sought.

The series covers systems and paradigms that employ knowledge and intelligence in a broad sense. Its scope is systems having embedded knowledge and intelligence, which may be applied to the solution of world problems in industry, the environment and the community. It also focusses on the knowledge-transfer methodologies and innovation strategies employed to make this happen effectively. The combination of intelligent systems tools and a broad range of applications introduces a need for a synergy of disciplines from science, technology, business and the humanities. The series will include conference proceedings, edited collections, monographs, handbooks, reference books, and other relevant types of book in areas of science and technology where smart systems and technologies can offer innovative solutions.

High quality content is an essential feature for all book proposals accepted for the series. It is expected that editors of all accepted volumes will ensure that contributions are subjected to an appropriate level of reviewing process and adhere to KES quality principles.

More information about this series at http://www.springer.com/series/8767

Sergey V. Zykov

Managing Software Crisis: A Smart Way to Enterprise Agility

Sergey V. Zykov
Higher School of Economics
National Research University
Moscow
Russia

ISSN 2190-3018 ISSN 2190-3026 (electronic)
Smart Innovation, Systems and Technologies
ISBN 978-3-030-08573-5 ISBN 978-3-319-77917-1 (eBook)
https://doi.org/10.1007/978-3-319-77917-1

Printed on acid-free paper

This Springer imprint is published by the registered company Springer International Publishing AG part of Springer Nature
The registered company address is: Gewerbestrasse 11, 6330 Cham, Switzerland

To God, my teachers, and my family

Foreword

As computers and machines have become increasingly intelligent and capable of performing complex activities, the software that lies at the heart of every action is under more pressure to deploy successfully each and every time it is called upon. Yet companies rush the production cycle in order to be first to market; machines are called upon to operate in environments for which they were not designed; and enterprises lose business with computer software crashes. The resulting unhappy customers, expensive opportunity costs, and potentially unsafe operations have led to a crisis in software production. Thus, this book is timely in addressing one of the most challenging problems in technology and one that is mostly hidden—developing reliable flexible software that works "out of the box."

Indeed, software development has been studied for decades. Project management techniques have been applied to the software development life cycle and have led to recommendations for methods such as waterfall, agile, object-oriented, scrum, lean, iterative. However, rather than focusing on a single method, this book takes a broader approach and investigates software production complexity resulting from the interplay between software quality characteristics, technological factors, and human-related factors. Issues and best practices for software development are illustrated with case studies.

This book will be a valuable and thought-provoking read for anyone interested in software development. The authors are experts who have studied both the problems and the successes associated with software. Their combined wisdom will benefit the community and hopefully contribute to better software in the future.

Baltimore, MD, USA

Dr. Gloria Phillips-Wren, Ph.D.
Professor and Chair
Department of Information Systems
Law and Operations
The Sellinger School of Business and Management
Loyola University Maryland

Acknowledgements

I would like to thank the colleagues of mine who essentially contributed to this book. They clarified initially vague ideas and helped with translation, copyediting, proofreading, diagramming, etc. Many of them are the students who did their master/Ph.D. theses under my supervision. Some of their papers findings and takeaways were transformed and included into this book as case studies on agility improvement and crisis responses. They are: Vlad Abdulmianov, Eunice Agyei, Artem Aslanyan, Vera Ermakova, Nikita Fomichyov, Ramis Gabeydulin, Prof. Alexander Gromoff, Alexandra Gureeva, Mikhail Kupriyanov, Maria Mamontova, Dinara Nikolaeva, Isheyemi Olufemi, Victor Rotari, Gaurav Sharma, Grigory Shilin, Alexander Sivtsov, and Sabina Supibek.

I would like to thank the Springer executive editor Dr. Thomas Ditzinger and the Springer project coordinator for books production Mr. Ayyasamy Gowrishankar, for their continuous availability and prompt assistance.

In addition, I would like to express my deep appreciation and sincere gratitude to the editors in chief of the Springer series in Smart Innovation, Systems and Technologies, Prof. Lakhmi C. Jain and Prof. Robert J. Howlett, for their cooperative efforts in supporting my initiative.

Contents

1 The Agile Way 1
1.1 Introduction: Adjustment for Agility 1
1.2 What Is Agility? 2
1.3 The Story of Russian Bridges 8
1.3.1 Why Is the Number of Bridges so Small in Russia? 8
1.3.2 Bridge Collapses in Russia 10
1.3.3 Building the Kerch Bridge in Crimea 11
1.4 Digital Transformation 11
1.4.1 Transparent Voting Platform Based on Blockchain 11
1.4.2 Decentralized Applications 16
1.5 Architecting for Agility 17
1.5.1 Sentiment Analysis System Based on Events Feedback . . . 17
1.6 Conclusion: How Agile Way Works 27
References 28

2 Agile Languages 35
2.1 Introduction: Communication Agility 35
2.2 Why Languages? 36
2.3 Making Processes Communicate 40
2.4 Developing an Embedded System 54
2.5 Conclusion: How Languages Work 62
References 63

3 Agile Services 65
3.1 Introduction: What Is an Agile Service? 65
3.2 The Microservice Approach 67
3.3 Enterprise Microservices 71
3.4 Bank Microservices 82
3.5 CRM Microservices 87
3.5.1 Analysis of Existing Solutions 89

3.6 Cloud Services 90
3.6.1 Process Modeling for Virtual Machines in Clouds 90
3.6.2 Information Process Model 92
3.6.3 Optimization of Virtual Machine Configuration 95
3.6.4 Experimental Design of Automatic Virtual Machine Configuration 97
3.6.5 Conditions, Objects and Order of Testing 98
3.6.6 Analysis of Testing Results 101
3.7 Conclusion: How Services Work 103
References 104

4 Agile Patterns and Practices 107
4.1 Introduction: Why Agile Patterns and Practices? 107
4.2 Agile Knowledge Transfer 108
4.3 Open Education Metadata Warehouse 118
4.4 Aircraft Communication System 125
4.5 Conclusion: How Patterns and Practices Work 132
References 133

Conclusion: Agility Revisited: What, Why and How 135

Glossary 139

Index 151

Acronyms

3D	Three dimension
AAC	Airline Administrative Control
ABS	Automated banking system
ACARS	Aircraft Communications Addressing and Reporting System
ACDM	Architecture-centric design method
AI	Artificial intelligence
AOC	Aeronautical operational control
APEC	Asia-Pacific Economic Cooperation
API	Application programming interface
APL	Application programming language
ARIS	Architecture of integrated information systems
AST	Abstract syntax tree
ATAM	Architecture tradeoff analysis method
ATC	Air traffic control
BPM	Business process modeling
CABS	Centralized automated banking system
CAD	Computer-aided design
CASE	Computer-aided software engineering
CD	Continuous delivery
CEO	Chief executive officer
CERN	Conseil Européen pour la Recherche Nucléaire
CI	Continuous integration
CID	Common interface definition
CIDL	Common interface definition language
CMU	Carnegie Mellon University
CORBA	Common Object Request Broker Architecture
CPU	Central processing unit
CQRS	Command Query Responsibility Separation
CRM	Customer relationship management

CSV	Comma-separated values
DB	Database
DBMS	Database management systems
DOM	Document Object Model
DPU	Data processing unit
DRE	Direct-recording electronic
DSL	Domain-specific language
DSM	Domain-specific modeling
DSML	Domain-specific modeling language
EA	Enterprise architecture
EAI	Enterprise application integration
EMF	Eclipse Modeling Framework
EPL	Eclipse Public License
EQCS	Education quality control system
ER	Enterprise resource
ERP	Enterprise resource planning
ETL	Extract–transform–load
FIFA	Federation International of Football Association
FTP	File Transfer Protocol
GDP	Gross domestic production
GMF	Graphical Modeling Framework
GRPC	Google Remote Procedure Call
GUI	Graphical user interface
HTML	Hypertext Markup Language
IAAS (aka IaaS)	Infrastructure as a service
IDE	Integrated development environment
IDL	Interface definition language
IEEE	Institute of Electrical and Electronics Engineers
IIOP	Internet Inter-ORB Protocol
IOT (aka IoT)	Internet of Things
IS	Information system
ISO	International Organization for Standardization
IU	Innopolis University
JDBC	Java Database Connectivity
JSON	JavaScript Object Notation
KAIST	South Korea Advanced Institute of Science and Technology
KPI	Key performance indicator
KT	Knowledge transfer
LED	Light-emitting diode
LISP	List processing
LMS	Learning management system
LOM	Learning Object Metadata

LOP	Language-oriented programming
LW	Language workbench
MEPHI (aka MEPhI)	Moscow Engineering and Physics Institute
METAR	Meteorological Aerodrome Report
MIT	Massachusetts Institute of Technology
ML	Machine learning
MODS	Metadata Object Description Schema
MOOC	Massive open online course
MS	Microsoft
NATO	North Atlantic Treaty Organization
NLP	Natural-language processing
NPP	Nuclear power plant
ODBC	Open Database Connectivity
OER	Open educational resources
OOOI	Out, Off, On, In
ORB	Object request broker
OS	Operating system
PAAS (aka PaaS)	Platform as a service
PDF	Portable Document Format
PL	Production life cycle
PLM	Production life cycle management
PR	Public relations
PROLOG	PROgramming in LOGic
R&D	Research & Development
RAM	Random-access memory
RDF	Resource Description Framework
REST	Representational State Transfer
RPC	Remote procedure call
RUR	Russian ruble
SAAS (aka SaaS)	Software as a service
SE	Software engineering
SML	Standard Meta Language
SMS	Short Message Service
SOA	Service-oriented architecture
SOAP	Simple Object Access Protocol
SQL	Structured query language
SSL	Secure Sockets Layer
SVM	Support vector machine
TCP/IP	Transmission Control Protocol/Internet Protocol
TF.IDF	Term Frequency–Inverse Document Frequency
UI	User interface
UML	Unified Modeling Language
VHF	Very high frequency
VLDB	Very large database

VM	Virtual machine
VS	Visual Studio
XML	Extensible Markup Language
XPATH (aka XPath)	XML Path Language
ZKP	Zero-knowledge proof

Abstract

This book is about enterprise agility and crisis-resistant software engineering.

Chapter 1 gives an overview of the concept of enterprise agility and its impact on organizational flexibility. This includes an in-depth evaluation of modern enterprise-scale projects and case studies on their non-agility, which often results in disastrous consequences. The ways we suggest to avoid these mission-critical errors are based on the existing research results in the field. Another key aspect is the agility improvement by means of the architectural trade-offs. We outline a few types of crisis recovery strategies and determine their links to the software architectures. We consider mission-critical systems for deeper understanding of how to increase scalability and avoid potential design errors. We provide case studies of bridge construction as examples of careless crisis management and foundations for agility improvement. We also focus on the digital integrity protection and particularly blockchain technology, and its application to secure voting. Afterward, we discuss decentralized applications and using their bit-torrent sharing networks in cryptocurrencies. We present a case study on sentiment analysis and its application to the agility issues in crisis.

Chapter 2 discusses the concept of programming languages and their use in application development. We analyze the evolution of programming languages from primitive to high-level ones. We examine the concept of domain-specific languages and a few problem domains for their testing and verification under specific environments, the focus being on inter-process communication with CORBA and ICE technologies. This chapter also describes the embedded systems and how they promote agility in mission-critical systems.

Chapter 3 describes the best practices in service-oriented enterprise software development. We define Microservices and illustrate how they improve organizational agility. We analyze how service-oriented architectures and particularly Microservices differ from the monolithic approach and identify their potential application areas. We discuss implementations of continuous delivery and continuous integration of the application life cycle and illustrate their importance. We present a case of Microservices for the banking sector and investigate how it helps to meet the requirements. We examine the integration of customer relationship management and geo-marketing and identify the business value of this synergy. We discuss the cloud services (and particularly virtual machines) as an agility booster; this includes in-depth testing of the proposed architecture.

Chapter 4 focuses on how the best software development practices depend on human-related factors; we further investigate the pattern-based approach as an agility driver. We discuss the principles of knowledge transfer and detect the key factors, which promote agile transfer in crisis. We look at the application of these factors to a Russian start-up, Innopolis University, and analyze the implications on the project flexibility improvement. We discuss patterns and anti-patterns of crisis-resistant development for increasing agility in mission-critical systems.

Keywords Agility · Architecture · Crisis resistance · Blockchain
Decentralized application · Life cycle model · Software product
CASE tool · Functional · Logical · Object-oriented · Domain-specific language
Reusability · Embedded system · Enterprise software development
Microservice · Customer relationship management · Banking · Cloud service
Best practice · Layer-based approach · Knowledge transfer · Design pattern

Introduction: Agility or Extinction

The focus of this book is smart agility management in crisis for large-scale software systems.

Marx explained crises and their nature. He stated that crises result from misbalanced production and the realization of a surplus value on the market [1]. The root cause of this misbalance is the separation between the producers and the means of production [2]. In software development, the nature of crises is somewhat different; a crisis is typically a disproportion between client's expectations and the actual product behavior.

Enterprise systems are typically complex; they combine a large number of hardware and software components. Managing the development of such complex software products, even under uncertainties and in crisis, is a key aim of software engineering. This discipline emerged in the 1960s as a response to the so-called software crisis. This term originated from the critical development complexity because of overwhelming computing power. In 1967, the issue became so critical that NATO had to arrange an invitation-only conference to find an immediate solution. The conference was held in Germany; its key participants were such gurus in computer science as A. Perlis, E. Dijkstra, F. Bauer, and P. Naur. These researchers and practitioners were Turing Award winners and the NATO Science Committee representatives from the USA, Germany, Denmark, and the Netherlands. At that time, the complexity of the hardware and software systems became unmanageable by the traditional methods and techniques. A large number of software products were late, over budget, or totally unsuccessful. To clarify this state of software production management, F. Bauer introduced the term "crisis" at the conference; E. Dijkstra used it later in his Turing Award lecture. The participants suggested "software engineering" (this term is also attributed to F. Bauer) as a remedy for that crisis. The idea was applying engineering methods of material production to the emerging large-scale software development in order to create better measurable and less uncertain products, i.e., to achieve a new agility level.

Now, we are on the same page concerning the crisis. However, what is agility? Intuitively, it is clear. For some experts, agile means flexible. For the other experts, agile means adjustable. Before giving a formal definition, let us turn to an example.

One can easily imagine a dancing puma or cat; these are typical instances of agile behavior or agility. However, can you imagine a waltzing elephant or a bigger (and more outdated) creature, like a mammoth or a dinosaur?

Why do we turn to large ancient beasts and waltzing? This is a *metaphor*. Metaphor is often used as an agile software engineering technique in order to disseminate a novel software system concept among the shareholders.

In our case, waltzing is an example of activity that requires agility. An elephant or a mammoth is a large-scale company which is an instance of a slow responsive and hard-to-adjust actor. The challenges of agile management are easier to track and monitor by means of large-scale company cases. An old-fashioned dinosaur example refers to an enterprise with complex and somewhat outdated management patterns and legacy computer systems. State-of-the-art business environment requires increased responsiveness, which is similar to following a tune while waltzing. Moreover, waltzing requires complex motion and constant coordination of the dancer's body parts.

Waltzing also requires a partner, which involves another level of coordination. This dancing style has a clearly recognizable pattern, which is certainly different from any other, such as salsa or polka. Though following a specific style might appear too difficult for an elephant or a mammoth, performing in any different style from the partner will likely result in a crash. This is due to an unexpected and/or unpredictable behavior, which in fact is an example of a crisis. Conversely, to perform harmoniously, i.e., to avoid a crisis, the partners need to coordinate (i.e., monitor and adjust) their actions following the same style. Therefore, to reach this agile harmony the partners need to master the waltzing process through training which requires both personal and team activities.

Currently, the Earth faces a global warming, which looks so slow that certain individuals do not notice it. As the warming accelerates, it may result dramatically for these non-responsive individuals in just a few decades. In crisis, it is very risky to remain old-fashioned and non-agile. Dramatic business climate change, uncertainties of resources, business and technical requirements, emerging and collapsing markets are radical and may result in critical consequences. In crisis, agility requires instant attention as insufficient agility is synonymous with extinction.

Agility is related to balancing business requirements and technology constraints. Many local crises result from misbalancing of these two aspects. Therefore, a well-balanced software solution means better agility. In other words, agility is a remedy for crisis. In crisis, agility is vital for any kind of a business structure. However, agility is better observed and understood in large-scale structures such as enterprises. Thus, building enterprise-scale systems requires a good balance. Not only should this balance be present in the design and construction, but also it should be present in each and every stage of the enterprise system life cycle. In the case of an enterprise, its agility is a concept that incorporates the ideas of "flexibility, balance, adaptability, and coordination under one umbrella" [3].

The early crisis in software development and the recent global economic crisis taught us a few lessons. One very important lesson is that the so-called human factor errors, which result from critical uncertainties and undisciplined life cycle management, often dramatically influence the software quality, and project success. Our systematic approach to the impact of these human-related factors on agile enterprise system development embraces the perspectives of business processes, data flows, and system interfaces. These three perspectives correspond to dynamic, static, and system architectural views. For each of the perspectives, we identify a set of business management levels, such as strategic decision-making or everyday management. After we combine these perspectives and the business management levels, we get the enterprise engineering matrix (Fig. 1).

BUSINESS PROCESSES	DATA FLOWS	SYSTEM TYPES
STRATEGY	STRATEGY	BI / PORTAL
INTEGRATION / KNOWLEDGE ANALYSIS	*METAKNOWLEDGE = = WISDOM*	*METAKNOWLEDGE = = WISDOM*
RELATIONSHIP MGMT	RELATIONSHIP MGMT	CRM / SCM
INTEGRATION / DATA ANALYSIS	*METADATA = = KNOWLEDGE*	*METADATA = = KNOWLEDGE*
RESOURCE PLANNING	RESOURCE PLANNING	ERP
PRODUCTION PLANNING	*SUPPLIES / ORDERS*	*SUPPLIES / ORDERS*
ACCOUNTING, DAILY MGMT	ACCOUNTING, DAILY MGMT	MES
PRODUCTION MGMT (PLANT LEVEL)	*TECHNOLOGY MAPS*	*TECHNOLOGY MAPS*
SUPERVISORY CONTROL	SUPERVISORY CONTROL	SCADA
TELEMETRY DATA COLLECTION/ HARDWARE DEVICE MGMT	*CLEAN DATA*	*CLEAN DATA*
DATA STORAGE	DATA STORAGE	DB / DWH
ANALOG-TO-DIGITAL	*RAW DATA*	*RAW DATA*
DEVICES/ SENSORS	DEVICES/ SENSORS	SENSOR / BOT

Fig. 1 Enterprise agility matrix

Another and perhaps a better name for this is the enterprise agility matrix as it determines enterprise agility. This matrix allows the detection of mission-critical dependencies in human (and other) factors for the systemic properties of the software products. These dependencies are based on relationships between certain values of the matrix cells. As such, we can build a set of constraints to guard the software development process from critical design errors and give an early warning of risky decisions. The matrix allows for agility estimation in terms of process management, data integrity, and interface quality. Informally, it will indicate whether the system set to be designed will behave like a puma or like a mammoth in an unstable, uncertain, or crisis environment. It will also recommend how to design an agile system that naturally accommodates to digital transformation.

This book is organized as follows. Chapter 1 covers the key concepts, such as agility and crisis, in more detail; it outlines crisis-resistant agility improvements for the age of digital transformation including blockchain-based decentralized software for transparent voting and a sentiment analysis application. Chapter 2 describes the history of programming languages and their agility in terms of large-scale software development; it discusses languages for domain-specific applications. Chapter 3 investigates the evolution of services in software development; it focuses on agile approaches including cloud computing and Microservices. Chapter 4 addresses agile and crisis-resistant pattern- and practice-based software development; it also discusses the human factors which promote software-related knowledge transfer. The conclusion summarizes the key outcomes of the book; it suggests agile ways to smartly manage software development in crises.

This book will recommend crisis adjustments and improvements in order to maintain agility and competitiveness in the global environment change. These recommendations are typically real case-based. Certainly, this book will not give a universal solution for agility. However, it will recommend technologies and approaches to start the new style of agile "dancing."

References

1. Chakravarty et al. (2013). Supply chain transformation: Evolving with emerging business paradigms. In *Springer Texts in Business and Economics* 2014th Edition, Kindle Edition.
2. Lowry, P. B., & Wilson, D. (2016). Creating agile organizations through IT: The influence of internal IT service perceptions on IT service quality and IT agility. *Journal of Strategic Information Systems (JSIS), 25*(3), 211–226. Available at SSRN: https://ssrn.com/abstract=2786236.
3. Chen, C., Liao, J., & Wen, P. (2014) Why does formal mentoring matter? The mediating role of psychological safety and the moderating role of power distance orientation in the Chinese context. *International Journal of Human Resource Management, 25*(8), 1112–1130.

Chapter 1
The Agile Way

Abstract This chapter gives an overview of the concept of enterprise agility and how this gives organizations more flexibility. This includes an in-depth evaluation of modern enterprises and a few case studies on how organizations failed to adopt agile principles which led to disastrous consequences, and ways to avoid their mistakes based on the research in this field. Another key area of this chapter deals with crisis resistant strategies and their links to enterprise architectures. Discussion of mission-critical systems is included in this chapter to allow further understanding of how to increase scalability and avoid potential design errors. Case studies in the bridge construction sector illustrate crisis management and agility improvement. The chapter also focuses on digital art integrity and blockchain technology application to secure voting. The other areas we discuss include decentralized applications and human-related sentiment analysis.

Keywords Agility · Architecture · Crisis resistance · Blockchain
Decentralized application

1.1 Introduction: Adjustment for Agility

This chapter presents the key concepts of agile management, specifically in crisis conditions and in large-scale systems. In the introduction to this book, we announced the basic concepts of agility and crisis. We also gave examples of the agile behavior in terms of animals (such as puma and mammoth), and gave a brief historical context for the crisis that triggered software engineering as a discipline.

However, these were intuitive definitions and a very quick sketch, just enough to get the preliminary understanding. For instance, some experts would state that agile means flexible. Others would argue that agile means adjustable. There is a vast space for more viewpoints.

Rather than tracing the complete history of agility or giving an exhaustive survey of crisis-related issues, the aim of this chapter is to suggest a deeper insight into the enterprise context of agility in crisis. That is why we started the introduction with the enterprise agility matrix, which embraces a systematic representation of these multi-

S. V. Zykov, *Managing Software Crisis: A Smart Way to Enterprise Agility*,
Smart Innovation, Systems and Technologies 92,
https://doi.org/10.1007/978-3-319-77917-1_1

aspect (i.e. process, data and component) views of hardware and software agility elements. Again, the right place to comment on these elements (at least some of them) and their intricate (and often subtle) relationships is the book itself rather than the introduction. Therefore, this chapter is to present a more detailed outlook and link these key concepts to the other chapters that contain domain specific elaboration such as: agile lifecycles, smart models and technologies, patterns and practices, human-related factors, intelligent services, resonant knowledge transfer and the supporting case studies.

As such, this chapter will recommend smart crisis adjustments and improvements in order to maintain agility and competitiveness in crisis. These recommendations are based on real life cases. Certainly, this chapter will not give a universal solution for agility. However, it will recommend technologies and approaches to start and improve an agile software development style. Emerging areas of cybersecurity, big data, and the Internet of Things, to name a few, give rise to the new smart business trends such as decentralized applications and blockchain [1].

Here is a brief overview of what would be discussed in the chapter. Section 1.2 provides a wider coverage of the agility principles. Section 1.3 presents a case study of crisis-related issues in complex systems based on bridge construction in Russia. Section 1.4 discusses another case study on digital transformation, based on the blockchain as a state-of-the-art example of a smart technology. Section 1.5 analyzes a case study on feedback-based sentiment analysis application. The conclusion summarizes the results of the chapter.

1.2 What Is Agility?

This section is largely influenced by Gromoff's findings [2]; it discusses the enterprise agility under the scope of 'Industry 4.0' and the related scopes such as stakeholders, staff and knowledge management.

Agility has a long history as a concept; let us refer to Latin root *agilis*, which means being able to move with quick easy grace or being mentally quick and resourceful. The first is associated with the puma or the cat, the second is hard to associate with a particular creature. This term was coined by Scheer and Nagel [3, 4] who stimulated the interest in agility of the enterprise and business processes in the early 1990s.

The framework of the section is based on logical construction of agility perception as a multi-dependent concept. Agility is a multi-dimensional concept represented in terms of semantic structures or taxonomical relationships.

Despite a lack of consensus on the definition of agility, certain characteristics are commonly detected. Information Systems (IS) scholars mention speed, sense and response, and IT capability as the key elements of agility. In a broad sense, *enterprise agility* refers to ability to manage unpredictable changes (i.e. crises) in uncertain business environment [5–7].

From this definition, agility requires strong sensing and responding. A degree to which an enterprise is able to sense environmental change and respond to it, business opportunities and threats, determines its agility. Building and maintaining these sense and response capabilities is resource consuming [8]. Therefore, the IS approach suggests carefully considering the context for business agility. As such, [9] provides a list of factors (including human, technical and contextual ones) that influence the enterprise agility.

Thus, agility can be interpreted as 'customization', 'adaptation', 'reflexivity', and 'flexibility'. The concept of enterprise agility essentially enables and facilitates the organizational flexibility [10, 11]. It generally refers to a company's ability to mobilize, deploy and manage IT and other organizational resources [12]. Currently, enterprises are increasingly dependent on smart IT in their efforts to be agile, as technology advances the breadth and quality of their knowledge and processes [13].

IS research specifies the following aspects of enterprise agility: (i) customer agility, (ii) partnering agility, and (iii) operational agility [4]. Of these three, operational agility is particularly essential; it reflects the enterprise process improvement in order to increase innovation degree and competitiveness [4, 12, 13]. With operational agility, enterprises can flexibly redesign or recreate their processes in order to be continuously evolving and competitive [4, 12].

Therewith, operational agility implies changes in enterprise business processes, which in turn require continuous changes in their models. This motivates and drives the development of agile enterprise modeling, which improves the static process models, and supports crisis responsiveness [14].

In our view, any business organization is a large-scale, adaptive system, which includes social (i.e. human factor-related) and technical subsystems. Since (i) any social system operates through communication, and (ii) organization is a social system, organization means *communication*. Thus, a social system comes into being only as a result of communication, i.e. as soon as its two independent sides establish a common goal and set up their communication "protocol", which is a *language* that these parties use to reach this goal.

Therewith, we focus on knowledge- and human-centered enterprise architecture (we describe these in a greater detail in Chap. 4). Human-centered design is critically important for Industry 4.0. The core aspects of such innovative socio-technical systems are typically specified in the following perspectives [15]:

1. Organizational: (de)centralized decision-making, process and unified information field transparency
2. Human: routine operations transfer from human to machine, improved services and control for humans
3. Technological: the boundaries for human and machine responsibilities.

These perspectives were discussed by a number of experts including [15–19]; however, a huge amount of professionals deeply involved in the Industry 4.0 neglected the "agility problem".

Typical example of the restrictions for successful IT implementation in business process modeling (BPM) lifecycle is an attempt to build a process model based on the results of interviews of its participants. As such, experts often treat themselves as subjects and therefore analyze the processes according to their own perception and ignore the subjectivity of the others. However, experts treat the processes as objects, and the interviewed humans as parts of these objects. Thus the human-related factors turn from subjects into objects. As a result, the developed interviewing software system would reject individuals as free and responsible actors.

Thus, our system development paradigm changes from 'subject—object' to 'subject—subject'. This change allows: (i) modeling ourselves, and (ii) observing this modeling as if we are the systems. This viewpoint is a *reflection*, which in our narrow sense, is a change of the spirit of oneself that follows the act of cognition. Concerning process manageability in crisis, we understand the processes as: (i) operations to achieve system's goals and (ii) interacting parts of system's control. These two aspects require cognitive modeling.

Any activity results from achieving a certain goal, which is either internal or external in terms of the system. Since being inside the system limits the scope and may corrupt modeling results, in order to recognize the goal we have to place it outside the system.

To build a model, the subject has to transfer to the reflective position, which is external to the former and future states of the observer. The knowledge obtained in this position is also reflective.

At this stage, activities of the subjects are *reflexive* and *reflecting*, which are different hierarchical levels with different knowledge (we describe the reflective cognitive knowledge transfer in a greater detail in Chap. 4). The roles of a subject in a process are limited to: (i) consumer and (ii) producer. The process quality depends on how a subject sustains these roles. In the extreme positions of consumer-only or producer-only, cooperation is impossible. Balanced, i.e. agile, reflexive position is optimal for process quality.

In this subject-oriented approach, the subjects themselves cooperate in their own behaviors and relations rather than abstract matter. The parts of the system become equal partners.

Our discussion on enterprise agility is based on the following definition: "*Agility* is the ability to manage and apply knowledge effectively".

Business process quality has a clear trade-off between efficiency and accuracy. When stakeholder-based human factors are not considered carefully, there is a little chance for BPM to rely on balanced decisions (particularly in case of incomplete information and/or insufficient time). Therewith, the operational risks increase dramatically, and this may result in a crisis.

Stakeholders are expected to be efficient and accurate at the same time; however, in reality due to human factors they often act differently. Therefore, in case the business process goes right, the management takes it for granted and seldom tries to analyze it. Only serious consequences may trigger investigation. Thus, we conclude that: "*Agility* is the ability of socio-technical system to be qualitative".

In the lifecycle of a socio-technical system, various stakeholders interact in multi-staged processes. These interactions between stakeholders are key crisis risks as they often critically influence agility.

Large-scale systems with multiple stakeholders and complex lifecycles feature miscellaneous risks of educational, cultural and other kinds of diversity. In such situation, agile management critically influences software production lifecycles.

Open collaboration between the stakeholders promotes adaptive responsiveness to human-related uncertainties; it clears a way for extracting and applying context-specific process information. Therewith, open collaboration turns inflexible process architecture into human factor-centered context-specific awareness, and adapts workspace to effectively implement the key stakeholders' goals.

Artificial Intelligence (AI) is an emerging trend; it incorporates methods, tools and techniques that enrich enterprise process models with the information which allows their adjusting on demand. As such, the key idea of AI is promoting enterprise agility by means of semantic analysis of business process information flow.

In the recent years, market conditions in certain industries were changing so quickly that many companies are still in search of smart agile approaches. To retain clients, they continuously customize their business and IT architectures based on the feedback from sales and marketing. However, it is unrealistic to manage the rapid business environment changes by traditional top-down approaches, particularly in large-scale enterprises. These top-down approaches are typically based on a set of models, which represent the key aspects of business and IT architecture.

Along with the market development, a critical factor of management changes is 'clip' mentality-based instability. As a result, traditional CRM methods are inapplicable to a large fragment of the market, and the related intra-organizational issues inhibit crisis-responsive agile management.

Currently, knowledge-intensive processes have a dynamic lifecycle. This was quite different in the 1990s when ARIS methodology came into being. As a result of execution variability, process changes occur at the operational layer. These changes are undetectable by enterprise BPM managers and stakeholders. Therefore, these changes are detected much later, often in the release of the software product. BPM becomes outdated as the modeling process is out of the scope of key daily management. As such, the employees do not reflect on process context changes. Consequently, the agility of the business process exceeds that of BPM system, and the process gets out-of-control.

Further, we present the outcomes of interviewing over seventy Russian companies in the BPM market. The main issues result from such human factors as incompetence and inertia.

Our innovative approach to enterprise modeling should solve these key issues:

- Detect business experts and process owners
- Maintain state of the BPM
- Trace changes in subject area that happen after external requirement changes
- Implement intra-process communication between key executive centers
- Develop 3D dynamic BPM with the metrics for workflow throughput

The classical models are *imperative* as objects and attributes are consistent and structured. This kind of modeling often neglects rapid environment changes typically addressed by waterfall lifecycle. The business process lifecycle has the following stages: identification, discovery, analysis, (re)design, implementation and monitoring.

We found an evolution trend from fixed hierarchies to flexible structures, manifested through active cooperation and knowledge transfer. This trend means increasing agility in the enterprise [20–22]. Therefore, agility becomes a concept that "incorporates the ideas of flexibility, balance, adaptability, and coordination under one umbrella" [23]. Two key directions to increase enterprise agility are improving services and communication.

Services shrink implementation time of IT projects as exhaustive interactions with customers become redundant, and IT expansion into the clouds accelerates. Service-oriented architecture (SOA) supports dynamic business process development by means of business services available for any particular domain.

These business process repositories, however, are often built on top of industry "best in breed" practices that impose rigid procedures and reference models for packaged applications (such as SAP). In this case business processes agility is hard to implement due to process inflexibility and fixed organizational structures. Optimizing BPM by a traditional workflow management system is complex and time consuming; moreover, the resulting solutions are often unsuitable for rapidly evolving processes.

The other options are reducing coordination/communication costs between process sides, and enhancing service orchestration and choreography. This requires intensive and consistent cooperation of the development team with the customer to establish collaborative business processes. An agile, self-organizing team incorporates "individuals who manage their own workload, shift work among themselves based on need and best fit, and participate in team decision-making" [24].

The key criteria of agile roadmap for business and architectural modeling are:

1. Cross-team collaboration, working environment that promotes efficiency, innovation and creativity
2. Iterative development, dynamic requirement elaboration, continuous interaction of self-organized workgroups consisting of various experts
3. Modifiability supported by knowledge management
4. Every team member (and customer) participates in business process optimization
5. Decisions are collaborative

We consider an enterprise as an adaptive socio-technical large-scale system [24]. As such, any organization includes at least two subsystems: social and technical. As a social system operates by means of messaging communication, 'organization *is* communication'.

Context-aware business modeling maintains the consistency in the changes of the actual processes and their models.

We identify two key types of changes:

1. Explicit changes: result from policies and procedures initiated by the process owners, approved by stakeholders, and implemented. Determined by negotiations with stakeholders
2. Implicit changes: result from occasional decisions of managers, and/or non-compliant behavior. Result in operational risks if unidentified. Manifest in omitted/redundant activities, outdated procedures/equipment, lack of staff, fatigue etc.

Information on enterprise architecture is often unstructured and represented as a silos of ad-hoc tasks and heterogeneous documents on processes, business policies, KPIs etc.

The information field is a set of messages the employees use daily and a set of internal regulations for their activities.

To synchronize enterprise models with the changes, we recommend the following steps:

1. Identifying changes

Learning phase: to create a dataset that maps the new context to model profiles, a taxonomy of models is built containing the model-related terms identified by experts. After the manual classification, the text analyzer processes all messages that belonged to each model profile. This service converts text for analysis by a linguistic processor; the results are used in Change Detection service and Expert Search service.

Discovery phase: the Change Detection service continuously monitors unstructured content to detect changes associated with new themes, events, and other objects. The results are used for change management. The service allows prediction of changes. This information gives a set of new terms associated with a model. Expert Search service handles exceptions and the cases that require higher level knowledge for decision-making.

2. Identifying experts

Expert search algorithm assesses and accesses tacit knowledge by linking it to the artifacts of explicit knowledge. It analyzes the documents associated with model in order to find evidence for expertise.

Experts search for people with high competence; they assume that personal qualifications correlate with the domain-specific set of characteristic concepts. Let us assume that among employees significant terms have a strong distribution of relative use frequency, whereas common ones have nearly identical frequency. We identify experts by a set of parameters including this relative frequency.

The Expert Search service returns a list of the most competent enterprise experts sorted by level of their knowledge on the subject.

3. Analyzing changes, adjusting model

The experts from the list produced at Step 2 analyze changes and adjust the model. The team members estimate labor volumes and decide on priorities. They are responsible for the change including decision-making on its importance and the activities required. Decision-making is based on a software system and daily meetings.

4. Modeling and Approval

Agile approach focuses on short iterations, clear deliverables and transparent communication. The Model Owner decides on actions and incremental model changes.

Architecture is paramount for organization development. The following aspects make *Enterprise Architecture* (EA): (i) technology, (ii) business concepts, (iii) organizational architecture, and (iv) information architecture. Each aspect encapsulates certain domains in sub-architectural layers; these should be coherent and consistent.

Business architecture and *SOA* are common terms; however, case studies on EA are few [25]. The business architecture concept is used in information management standards (IEEE 1471, ISO 15704 [26, 27]), classifications [28, 29] and strategic enterprise management by IT solutions (IBM, Ernst and Young). This is why EA links to IT.

According to Hoogervorst, EA includes: (i) business architecture, (ii) information architecture, (iii) technology architecture and (iv) organizational structure [30]. EA separates the business and technological aspects of complexity by means of layering.

Gromoff found that business architecture is a frame of operations realized reflectively by self-organized groups of business subjects. Therewith, business subject should be distinguished from routine business resources by internal motivation to reach the business goals [31].

Architectural layering starts by separating the business and the system models, and forming system design requirements. Further, this layering separates logical and physical design. For traceability between the conceptual, logical and physical layers, architects use the concepts of *business requirements* and *design specifications*. Business requirements identify the constraints for a technology solution to support a business; design specifications describe this technology solution to meet the requirements.

1.3 The Story of Russian Bridges

1.3.1 Why Is the Number of Bridges so Small in Russia?

The 1968 NATO Software Engineering conference used a bridge as a typical instance to sample a complex software system, and bridge development and maintenance to sample lifecycles in software engineering [32]. A bridge itself works in multiple interacting environments (such as water and air); and it should possess certain qualities (maintainability, reliability, throughput etc.). Bridge construction typically requires several interacting teams specializing in different areas.

The construction of Russia's biggest Kerch Strait bridge between Crimea and the Russian mainland is likely to provide some ground to discuss the crisis cases in software development, and relate these development crises to human factors.

Apparently, a company called "Stroygazmontazh" that had no relation to the construction of bridges previously, has received the contract to build the important crossing without competition.

According to Rosstat, there were 42,000 bridges and overpasses in Russia by the end of 2014. This is only 200 units more than in the year 2000. The length of all the structures in that time has grown from 1.5 million meters to 2.1 million, whereas the length of the roads over the same period has increased by 70% [33].

There are no official statistics about the lack of bridges in Russia, however there are two indicators. First, it is the number of constantly functioning crossings, both in summer and winter. There are 257 summer and 3,500 winter crossings, which is a very high indicator, even for such a large country.

Second, enormous transport rerun (mileage in comparison to the actual distance between two destinations) can reveal the lack of bridges. "In Moscow, the rerun is 70–80% against 20–25% in Western metropolitans", said Probok.net and there is no city in Russia with an index better than 50.

Every ninth bridge in Russia is made of wood. Over the past 15 years, their total length was reduced by approximately 30%.

However, wooden bridges are not necessarily bad: in some regions, it is a cheap and perhaps the only replacement for reinforced concrete bridges, which are unlikely to be built. What is the state of most of the bridges in Russia?

The part of federal roads (including bridges) that meet regulatory requirements in 2015 was slightly more than 60%, according to the Rosavtodor. According to Global Competitiveness Report, Russia rated 123 out of 140 in the quality of roads, and located between Sierra Leone and Benin.

Around 20% of bridges in Russia in the early 2000s were assessed as "good" by the Ministry of Transport. Some of the bridges were unsatisfactory and around 1% of them were emergency. Over 500 bridges in Russia were unsafe. By 2013, the number of safe bridges increased by almost a hundred.

Since 1995, the Ministry of Emergency Situations has recorded 26 automobile bridge collapses.

The expenditure on the roads in Russia is extremely small in comparison with other countries, measured as a percentage of GDP, is 179% in Japan, 76% in China, and 60% in Russia.

The Russian Transport Strategy signed in 2008 reports that more than 10% of the population, 15 million people in the spring and autumn remain cut off from transportation and communications because of the lack of paved roads; about 46,600 settlements have no connection to the transport network through the country's paved roads; there is a lack of basic network of roads in the districts of the North, Siberia and the Far East; more than 30% of roads are congested.

According to SPARK, there are approximately 500 companies in Russia with total revenue over 50 million rubles, which construct bridges and tunnels. In 2014, their aggregate revenues amounted to 638 billion rubles, where about 20% belonged only to Mostotrest. In 2014, this company earned 119 billion rubles.

Surprisingly, top companies by the number of contracts for the construction of bridges, compiled by the Committee of civil initiatives are very different from the top list by revenue.

Construction costs are increased by the traditional Russian difficulties: weather, distance, and urbanization of the road network.

Russian bridge building is critically dependent on imports. For example, the cables for the bridge on the Russky Island (Far East), the basis of the entire design, was developed and installed by the French company Freyssinet. French cables were painted afterwards in Russian tricolor. Production of own cables in Russia is minor.

1.3.2 Bridge Collapses in Russia

Saving on investment in infrastructure sometimes leads to emergencies. For example, four bridges collapsed in 2015–2017 only in Primorsky Krai (Far East) [30]. Most of the bridges in the region that hosted the APEC summit in 2012 had not been repaired in about 30 years, the administration of the province said. Even now people are suffering or dying from the collapse of bridges in Russia.

At the end of 2015, one after another, three bridges fell in the Arkhangelsk region. Firstly, on October 20, a bridge built in 1966 across the Vaga River in the Velsky district collapsed. 12,000 people remained without automobile communication, 200 children could not attend school. According to the "Arkhangelskavtodor", the last time the bridge was repaired, was in 1987.

A wooden bridge over the Solymka River was destroyed in May 2015 by a 40 ton dump truck. There were no signs before the bridge prohibiting the passage of heavy vehicles, therefore, it was impossible to bring the driver to justice. But, according to the Head of the Kosinsky district, this driver "should understand that the bridge is wooden and cannot withstand such a machine".

February 23, 2015, the 1965 bridge over the Shurap River collapsed in the city Berezovsky of Kemerovo region. No one was hurt. The reason for the collapse of the building was given as structural fatigue.

The victims of the collapse of the Palmburg, or Berlin, bridge in Kaliningrad on January 13, 2015, were four workers. The old bridge, built in 1938, began to be dismantled in December 2014. According to the investigation, the reason for the collapse of the bridge was safety violations, as the bridge could not stand heavy equipment.

Four bridges in Primorye collapsed in three months in 2016. In addition, on February 23, along with a passing car two spans of the bridge in the Litovka River collapsed. Fortunately, no one was hurt. The damage to the regional budget was estimated at 30 million rubles.

1.3.3 *Building the Kerch Bridge in Crimea*

The construction of the bridge started in 2012, it will cost 228.3 billion rubles [31]. There are no other buildings in Russia of comparable cost. For example, the bridge to Russian Island in Vladivostok that is longer than 3 km cost 35.4 billion rubles. The bridge across the Golden Horn Bay (1.4 km) cost 20.1 billion rubles. The cost of a bridge across Lena River that has not been built was estimated at 54.9 billion rubles.

President Vladimir Putin said that it was necessary to build an automobile and railway bridge across the Kerch Strait, immediately after the Crimea joined Russia in 2014. Currently the possible ways to arrive to Crimea from the mainland of Russia, bypassing Ukraine, are an airplane or a ferry. In 2015, the ferry across the Kerch Strait was used by 5 million people.

The length of the arched span of the bridge will be 227 m, the height of the fairway arch will be 35 m. The bridge will be constructed concurrently in six sections to withstand the necessary rate of work and to meet the deadlines. In 2016, the builders erected the first support of the bridge under the motorway on the Tuzla Island. More piers are under construction.

In 2016, the cargo ship Lira, belonging to Turkish Turkuaz Shipping Corp., destroyed one of the supports of the bridge. The damage was estimated at 120 million rubles.

The bridge should start working in December 2018. However, Reuters reported that the construction of the railway bridge across the Kerch Strait to the Crimea will be completed later than scheduled. The construction of the railway bridge requires more time and is technically more complex than that of the motorway one. The railway segment will not be launched until the end of 2019.

1.4 Digital Transformation

1.4.1 *Transparent Voting Platform Based on Blockchain*

During the 2010s, intensive research has been focused on the challenges in centralized voting systems, e-voting protocols and decentralized voting. Electronic voting puts forward certain difficulties regarding the voter anonymity, the secure casting of votes and preventing fraud. Decentralized blockchain technology could help solving a number of these challenges as it provides a new, secure, safe and transparent voting mechanism.

Let us find the balance between voter privacy and transparency.

The integrity of digital information requires *blockchain* technology, a decentralized and distributed database in a peer-to-peer (p2p) network. In a blockchain system the data is shared between the nodes of the p2p network. Transactions in a blockchain system are public. If a user tries to make a transaction (sending, receiving Bitcoins or

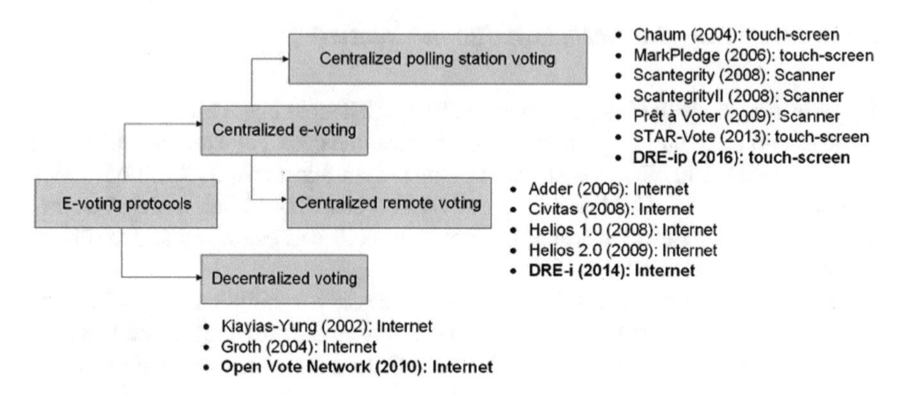

Fig. 1.1 E-voting systems

casting a vote), the system verifies the transaction before adding it to the blockchain. This verification prevents the double spending or the invalid votes.

In other words, the *blockchain* can be defined as a list or a decentralized ledger of all the transactions that are processed in a p2p network. Blockchain technology is used in Bitcoin and other cryptocurrencies.

Fraud includes double voting, buying the votes and using the blank ballots. In paper voting, there is always a trusted party responsible for counting the votes. Since only the trusted party does the verification, the voters cannot find a clear way to check and verify the final results.

Typewise, Chaum [34] proposed the first voting system based on cryptographic and mix protocol (Fig. 1.1). The centralized remote voting systems include Civitas [35], DRE-i [36], Adder [37] and Helios [38], to name a few. The other voting systems based on polling stations are MarkPledge [39], Prêt à Voter [40], Votegrity [41], DRE-ip [42], STAR-vote [43] and Scantegrity [44]. Some of the decentralized systems are Groth [45] and Kiayias-Yung [46]. Of the above, the only systems without tallying authority are DRE-ip and DRE-i.

In 2016, the Economist and Kaspersky Lab organized a challenge on blockchain-based secure digital voting between 20 universities [47].

The Votebook system (New York University) won the competition. It allows a centralized authority to be responsible for the way that the encryption keys are distributed on the network nodes; this requires a permission blockchain. Each node in the network is a voting machine. Every voting machine generates public and private keys. The private key is stored securely, and the public key is sent to a centralized authority.

Every network block contains: (i) node identifier; (ii) timestamp; (iii) hash of the previous block and (iv) set of voters with votes and digital signatures for their validation process. Some of agile design considerations were: verifiability, no coerced voting, publishing/hiding interim results on demand, allowing voter abstinence, and minimum behavioral changes. The challenges were e-voting vulnerabilities, such as:

Sybil/DOS attacks, voting machine tampering, too abstract algorithms for hashing and private/public keys generation, and anonymous identity verification.

The Newcastle University team proposed a decentralized *Open Vote Network* system where the trusted authorities were removed from the process of the election. The proposal was focused on the possibility of having electronic voting protocols with the use of Ethereum blockchain as a self-enforcing system.

It cast the votes on the distributed p2p network in multiple rounds, and the voters verified the last tally in a private way. The system was decentralized and used a voting scheme of two rounds [48, 49]. Tallying gave the privilege to each voter to tally votes (also referred to as "self-tallying"). The proof of concept was Ethereum blockchain-based. Two smart contracts were implemented: a voting contract and a cryptography contract. The system included five election stages: setup, signup, commit, vote and tally.

The system was massive and bulky for the Ethereum blockchain-based computations as the language did not support elliptic curve cryptography.

The University of Maryland used the Ethereum blockchain to record the votes with the use of Zero Knowledge Proof (ZKP) and Merkle tree as cryptographic primitives. The Merkle tree proves to the voter that the vote is included in the counting process after the end of the elections. The ZKP proves the correctness of the tally process. In their system, each voting machine represented a voter with a server that was responsible for handling the decryption and the tallying process. The voter client here encrypted the vote with the public key of a centralized authority so that the authority handled the decryption and tallying in a correct and verifiable way. The proposal did not use a cryptographically approach to the cast of the vote, however it used a random number as a receipt.

The smart contract in their proposal was tallying every vote with spending coins in the voting process while choosing the candidate [50]. However, there was no guarantee whether it was possible to verify if the vote was cast as desired.

Concordia University suggested voting under unconditional integrity and privacy. This is a system with different security properties and interdependent combinations of security issues. The system depends on Eperio, which lets the voters cast paper ballots and is not involved in the tallying process [51]. A possible drawback is that the voter is not directly involved in the process of tallying.

The general attributes of a voting system are as follows. Integrity: the condition of the system to be unified should be always guaranteed [52]. Eligibility: only eligible voters can cast a vote. Availability: system must remain available in real-time during the process of elections. Fairness: the system should not publish any partial results before the time of voting ended. Anonymity: the voter identity should not been known except of the voter himself. Correctness: the final results of the elections must be counted correctly to be published. Verification: the system must verify the election results. Robustness: The system must handle ineligible votes and the votes which cause faults. The Concern of Coercion: the system must ensure that the user cast voice without giving vote to a specific candidate by force, and not let the user show vote to anyone else [53].

The cryptographic primitives include the public and private keys to manage the voter privacy. Each voter has one public key and one private key. Every voter uses the public key, which is in the election's public key, to encrypt his vote or to encrypt the ballot. This key is accessible to everyone. Then the voter uses the private key to sign the ballot which is already encrypted by the public key [54]. Mix net technique guarantees voter anonymity [55]. According to this technique, the voting system removes a layer of the encryption and then shuffles the order of the votes and sends the result to the next node after the votes have been encrypted [56–59]. To guarantee voter anonymity there should be at least one mix net server [60]. This mix net property includes one or more mixes which are connected to each other with a specific cascading order so that the outputs of one mix net make the inputs for the next one; this provides anonymous voting [61].

In 1981, Chaum came up with cryptography-based secure voting. He used a technique of public and private keys, which makes the participants identity unknown and hidden for public communication. However, this technique is not secure enough for real world elections [34, 62]. To maintain the privacy, the voter confirms that the vote was correctly encrypted by tracking it with a receipt. The voting system then publishes the cryptographic proofs of the correctness of the operation to ensure results integrity [61, 63, 64]. The verifiability in the end-to-end voting systems typically has three main steps:

1. Cast the vote as planned (one who voted can have the right to verify that his/her choice of the candidate on the ballot was correctly marked in the system)
2. Record the vote as casted (the voter can check if the system has recorded his/her vote correctly)
3. The vote tallied as it was recorded (the voter can check if the system counts the vote as recorded).

Zero Knowledge Proof In 1980, Goldwasser, Rackoff and Micali proposed the concept of *zero knowledge* [65]. The research area was "interactive proof systems" which explains how two parties (prover and verifier) can exchange messages in order to make the verifier agree that a specific mathematical statement is true. ZKP must possess the properties of completeness, soundness and zero knowledge. The latter means that the verifier will not get any knowledge and no information will be gathered except that the truth of the statement and this property represents the actual meaning for the proof zero knowledge [66, 67]. Non-interactive ZKP is used by most of voting systems as there is no need for two active parties in the system [68]. We recommend non-interactive ZKPs for use in voting systems as the voter is able to verify a number of steps without being an active part of the system, so this is resource efficient and agile [68].

The proposed architecture of an agile voting system To become agile, we recommend improving the Waves platform. This is a blockchain decentralized open source-based platform. It is fully functional for the transfer, issue and exchange of different cryptocurrencies (e.g. Bitcoin and Ethereum). The platform is auditable, decentralized and transparent. There is no need for the end user to download the entire blockchain, which provides accessibility.

Currently, maximum Waves block size is 100 transactions, and the generation speed is up to 100 block/transaction per minute. Every block can handle around 100 transactions. Transaction fees can be treated as tokens.

The system should have the following key features. First, it should provide voters privacy and vote checking/verification. Second, it should be able to remove any indication of the voter's identity. Third, unfinished voting results should not be recorded. Fourth, the voter should always vote securely. Fifth, blank votes cannot be used to elect a candidate or make a decision. Sixth, to maintain blockchain integrity the system should incorporate a number of active servers in different locations. Seventh, the system should use efficient and secure algorithms for key generation and encoding/decoding.

Let us outline high level system use cases:

1. Registration. Each government, which wants to create an election, and each eligible voter will create their own private wave wallet, and the system will verify their eligibility
2. Creating the issue/matter for voting. The system will set up the election process details and define the main area of voting. Addresses of specific polls/answers will be available publicly
3. Voting transactions process. For each answer, the voter will spend a certain amount of Waves, assets or currency depending on the poll requirements
4. Verifiability. The system will verify the user by means of voter ID/password and voter list. Each user can see if the vote is in the candidate's wallet. To reconstruct the results of the election, the other transactions will be verified similarly. Each candidate will specify a Bitcoin/Wave address. Voters then will cast by sending a payment to the selected candidate. Anyone trying to break the voting rules (e.g., one vote per voter) will be detected by inspecting the blockchain, and the tally will be visible by inspecting the candidate's received payments. Votes will be stored in packages with a predefined maximum size and verified with a hash

We suggest the following extension of the system use cases:

1. Coin replacement by vote. In a proof-of-stake (POS) system, the holding tokens (or Waves) will be alternative to hashing. The tokens will be transferred in the network describing individual votes by transferring them into ballots. For each voter, the system will vote for right assets and token assets. To cast a vote on each agenda item, a voter will spend a certain amount of voting tokens
2. Double voting prevention. To prevent double voting, before adding the transactions to blockchain, their inputs will be checked to verify that they were not voted previously. The protocol will define that the longest chain is the "true" chain; it will ignore shorter chains. Together with a timestamp and the POS, this will prevent double voting
3. Voting and creating elections as transactions. Each election/vote transaction will perform only one function, the record of which will be permanently stored after this transaction was included in a block. Election/vote transaction will be the primary mechanism through which the wave recirculates back to the network.

Each election/vote transaction requires a minimum fee of one Wave. All Waves election/vote transactions will be unconfirmed until included in a valid network block. Each election/vote transaction contains a deadline parameter

4. Transaction parameters and types. The key transaction parameters will be: a private key for the sending account, transaction fee, transaction deadline, and referenced transaction (optional). The transaction types will include subtypes supported by Waves. Each type will specify transaction parameters and processing method (i.e. messaging, poll creation and vote casting). If the sending account has funds for creating a election/vote transaction: (i) a new transaction will be created with the required parameters including a unique transaction ID; (ii) the transaction will be signed by sending account's private key; (iii) the encrypted transaction data will be packed in a message that will instruct the network peers to process the transaction; (iv) the transaction will be broadcasted to all network peers if successful; the server will respond with a result code (either election/vote transaction ID if the transaction was successful or error message otherwise)
5. Encryption. For security/performance balance, election/vote transaction data will be encrypted by Elliptic-Curve Korean Certificate based Digital Signature Algorithm (EC-KCDSA)

Based on our comparison of e-voting systems, we outlined the new Ethereum blockchain-based architecture. This architecture uses an agile Waves platform and a hash-based POS protocol which together provide a balanced security/performance tradeoff for real world solutions. The approach provides blockchain integrity for the p2p network elections with portable devices (such as smartphones etc.). Further improvement plans include the implementation of the proposed architecture on a Wave light client to provide a real voting product which will focus on elections scaled up to a national election. To create this type of large voting system, a dedicated blockchain will be exclusively responsible for voting; it will feature a bigger block size to handle a larger number of transactions.

1.4.2 Decentralized Applications

Decentralized applications have become increasingly important as they assist in massively reducing costs, preventing censorship and ensuring transparency and trust between all parties involved in interactions [69].

Bit-torrent, a file sharing network, developed in early 2000 is arguably the first decentralized application created. Bit-torrent allows sharing files and distributing content quickly and easily. Five years later, Nakamoto came up with the idea of a blockchain as sort of distributed database [70]; he proposed using that to build Bitcoin, world's first decentralized currency. Decentralized currencies allow sending money instantly anywhere irrespective of national borders at negligible fees. Bitcoin is increasingly being used for international remittances, micropayments and commerce online. Decentralized financial applications and distributed governance based on cloud computing are probably the "next big things" to come [71].

Ethereum is a platform to build these decentralized applications (Dapps). The Ethereum clients will include a built-in peer-to-peer network for sending messages and a generalized blockchain with a built-in programming language [72].

Allowing people to use the blockchain for future centralized applications, Ethereum can be used to build trustworthy and transparent financial software. Online cryptography systems allow security by managing properties in contracts. Social networking and messaging systems allow users to control their data. Systems for treating underutilized computational resources (e.g. CPU time and hard drive space), and rapidly evolving tools for online voting and distributed governance can also be Ethereum-based.

A *smart contract* is a program that runs on the blockchain and has its correct execution enforced by the consensus protocol [73]. A smart contract is a piece of software that stores rules for negotiating the terms of contracts, automatically verifies the contract and then executes the agreed terms. A smart contract is identified by an address (typically a 160-bit identifier) and its code resides on the blockchain. Users invoke a smart contract in any of the cryptocurrencies available by sending transactions to the contract address [74].

Currently, a number of countries support cashless economy and transactions. With the advent of Bitcoin and Ethereum blockchain, applications eliminated the third party previously required in the payments and introduced a new concept of customized currency. The UN plans to conquer the hunger in Jordan by introducing Ethereum blockchain-based coupons instead of the local currency [75]. Along with IRIS scan technology these coupons will help to distribute foods to local people.

1.5 Architecting for Agility

Let us see how sentiment analysis can apply to the agility issues in crisis. We give more details in Chap. 4 while discussing the human factor related challenges and possible ways of managing them.

1.5.1 Sentiment Analysis System Based on Events Feedback

This section focuses on social networks; however, it can easily expand to cover the cloud computing and other kinds of services (such as Microservices); we provide more details on services in Chap. 3.

A *sentiment* is an attitude, an opinion or a feeling toward a person, organization or product. *Sentiment analysis* is the process of using natural language processing (NLP), statistics, and/or machine learning (ML) methods to extract, identify, and characterize the sentiment content of a text unit [76]. The prevalence of networks (including social media, intranets i.e. internal enterprise networks, and extranets i.e. enterprise partner networks) has provided a means for employees (and people in general) to efficiently share information on the (project) activities they are engaged in.

Sentiment analysis, also known as opinion mining has been in existence long prior to the World Wide Web [77]. Enterprise activities (and human lives in general) feature intensive and complex decision-making. A strong tendency is to inquire from the experts (including enterprise employees), seek their opinions and recommendations in the decision-making processes. In crisis, the survival of businesses largely depends on decision-making: whether to enter a market with a new product now or later, how to price products or services or both etc. A typical example is increasing customer satisfaction. In their quest to satisfy and retain their customers, the enterprises conduct surveys to collect the opinions of customers. The feedback of these surveys is analyzed to make much more informed decisions.

The emergence of the World Wide Web has provided a platform for people to air their opinions on various websites and blogs. Later, Web 2.0, a form of World Wide Web emerged, placing a lot of emphasis on social networks [78], thus providing a medium for individuals to make their opinions known on various social media platforms without having to wait for a poll or survey. This prompted enterprises to use data from the social and other networks in order to make much more informed decisions. These enterprises realized that the opinions of customers could influence the decision of other customers [77].

The cognitive lifecycle for any event includes evaluation. Typically, events are evaluated using the traditional quantitative or qualitative approaches to collect and analyze data. These approaches are used to evaluate the success or failure of events and to make informed decisions on how to improve future events (we address cognitive lifecycle in Chap. 4). These approaches commonly include questionnaires, interviews, focus groups, etc. to get feedback on how well the event satisfied the attendees. This improves agility and makes the enterprise more crisis resistant. With the advent of social media, customers evaluate products and services by commenting on them on Twitter, Facebook, and the like. This infiltration of social media into the clients and enterprise employees generates a large data bulk as people share their opinions across the different platforms. As of January 2016, Twitter recorded an average of 303 million tweets per day [79]. This data burden has not been fully explored for event evaluation. Our case study focuses on the sentiment analysis application to improve the quality of cultural event organization. The case study uses tweets collected in Twitter at the Beale Street Music Festival (Memphis, Tennessee, the USA).

Culture portrays the identity of an individual with regards to customs, beliefs, religion and social values; giving an individual a sense of belonging and therefore, there is a need to preserve culture as there are tendencies to lose it especially in this era of modernization. The best way to preserve culture is to share with others [80]. Microsoft has developed an online solution for the indigenous people of Arhuaco to help them preserve their culture. This online solution houses a database of the records of the people, lands and sacred places [81]. Also, IBM has collaborated with Egypt, Russia, Italy, and United States among other nations to preserve cultural heritage. The External Egypt project is an example of how IBM is using technology to help preserve and present the Egyptian culture to the world. The prominence of social networks provides a good medium to reach out to the world to share culture in an attempt to preserve culture. Sharing on social media brings the possibility of feedback from the world. Sentiment analysis on this feedback can be used to improve some aspects of cultural events. Cultural events range from ceremonies and celebrations, rituals and rites of passage, to public performances and interacting arts. Organizing cultural events requires proper planning and implementation to attract people [82]. Cultural events are attended for many reasons including relaxation, socialization, event excitement and experience [83]. Attendees of cultural events bring revenue to the community; it is desirable that their expectations are met.

We outline the data collection from social networks and its preparation for sentiment analysis. As such, we use a specific architecture and a multi-level analysis process for the sentiment-based evaluation of cultural events. We perform sentiment analysis at word, sentence, document, and aspect level of the tweets.

A major problem of qualitative evaluation is the selection of participants who can affect the quality of the evaluation. Collecting data from social networks is random as one does not know participants. Hence social network data is a bias free medium to collect data, evaluating a certain event with sentiment analysis can help to adjust/improve the quality of future events, and thus to increase agility and crisis resistance level.

Tweet analysis typically requires using Application Programming Interface (API) by Twitter that can handle a large stream of data. We will analyze English tweets only as this is the most widely used language. We limit the problem domain to cultural events, and focus on the Beale Street Music Festival specifically.

A number of approaches and algorithms have been suggested for sentiment analysis; these can be primarily categorized into (i) machine learning (ML) and (ii) semantically-oriented.

The ML approach may be supervised or unsupervised. The *supervised* machine learning approach requires part of the data to have a sentiment label known as the training data set. This data is used to train a classifier model. Afterwards, remaining data is used to test the model [84]. Supervised learning techniques include Naïve Bayes Classifier, Support Vector Machine (SVM), and Max Entropy. *Unsupervised* machine learning does not require any training data. The algorithm classifies the data based on some distance measures like the Euclidean Distance. An example of unsupervised machine learning algorithms is K-nearest neighbor.

Naïve Bayes is a probabilistic classifier that classifies the polarity of words in a text using the Bayes theorem. This theorem treats each word in a text independently. Each word contributes to the probability that the sentiment is positive or negative independently without regarding the possible correlations between the words. *SVM* is a classification algorithm used to separate data into two clusters. Using a training dataset with a class label, SVM builds a model that assigns a new instance to one of the clusters. SVM relies on a labeled dataset to classify the polarity of text. Depending on the variant of Naïve Bayes or SVM, dataset, the task to be performed etc., one algorithm may outperform the other [85, 86].

Max Entropy is a probabilistic classifier based on the Principle of Maximum Entropy (PME). PME selects the most uniform distribution (with maximum entropy) from a set of possible distributions that satisfies the given constraint. This algorithm relies on dependencies between the features which is in contrast to Naïve Bayes algorithm. Max Entropy is based on the probability that a word, sentence or a document belongs to given a context must maximize the entropy of the system. It ensures that no biases are introduced in classification by maximizing entropy [87].

The *K-nearest* neighbor algorithm is also used to classify the polarity of words in text. It stores all available words and their labels. It then calculates the similarity between a new word and already labeled words using similarity measures like Euclidean Distance. It considers the label of K similar words and then assigns a label to the new word using the label of the nearest K-neighbor with a majority vote.

The *semantic orientation* approach performs sentiment classification based on the positive and negative sentiment of words in the text. The idea of semantic orientation of a word indicates how a word deviates from its semantic group [88]. Techniques used for this approach include: corpus-based technique, dictionary-based approach or lexicon-based approach. The *corpus-based technique* determines text sentiment by finding the co-occurrence of patterns of words. Turney classified the sentiment of a review by calculating the average semantic orientation of all adverbs and adjectives in a text [89]. *Lexicon-based approach* relies on lexical resources such as Sentiwordnet, MPQA-wordnet, WordNet-Affect proposed in [90–92]. These lexical resources contain items with polarity information [93]. It assumes that any text can be assigned a polarity based on the polarity of words. This approach may fail due to the complexity of natural language [94]. The *Rule-based* approach scans text, identifies opinionated words, and classifies them based on the number of positive and negative words using booster words, idioms, emoticons among other classification rules [95]. Statistical models classify opinions into negative and positive based on standard word sequences [96]. Sentences are scored based on the N-gram probabilities. An N-gram is a sequence of N words. The probabilities are estimated using frequency counts [97].

Earlier works demonstrate how sentiment analysis has evolved from a coarse- to a fine-grained analysis [98]. Sentiment analysis can be performed at the document level, the sentence level or the aspect and feature level. At the document level, the whole document is classified as positive, negative or neutral. This level of sentiment analysis assumes that a document expresses an opinion about an entity, hence it is not suitable for documents with comments on multiple entities [99]. A typical example of an entity with multiple features is an emoticon. The features of emoticons cannot be extracted for sentiment analysis [100]. At the sentence level, sentences are classified as negative, positive or neutral while aspect or feature level classifies a whole document or sentences as positive, negative or neutral based on certain aspects of the document or sentence. Sentence level and document level sentiment analysis is used mostly for micro blog analysis [101]. Sentence level sentiment analysis treats a sentence as an independent entity that carries an opinion [102]. Subjectivity classifications classify sentences into two groups; subjective and objective sentences. An objective sentence expresses factual information about the world, a subjective sentence conveys the view, opinions and feelings of a person [100]. Objective sentences may be subjective as they present both facts and the opinion of the writer. Aspect-based sentiment analysis identifies the aspects of a target entity and assigns a certain polarity to each aspect [103]. Aspect-based sentiment analysis is word or phrase level [101]. It is used to further understand the exact aspect of each positive or negative entity [104]. This level of sentiment analysis involves three main tasks: (i) identifying and extracting an entity from any given text, (ii) identifying and extracting the aspects of the entity, and (iii) determining sentiment polarities on both the entity and its aspects [105].

Nguyen et al. conducted public sentiment analysis using machine learning and dictionary approaches [104]. They mentioned that the major difference between the two approaches is that the machine learning approach classifies entire tweets whereas the dictionary approach classifies individual words. They collected tweets about the royal birth of 2013 using the Twitter Streaming API. According to them, machine learning is not used for aggregating public sentiment; however, machine learning can be used for aggregating public sentiment.

Relations among social media users can be used to improve sentiment analysis. The data collected from Twitter proved that connected individuals have similar opinions. The likelihood that two connected individuals hold the same opinion was higher than that for the non-connected ones [106].

Using a movie review data set, four feature selection schemes were generated from the dataset. These were: Term Occurrence, Term Frequency, Binary term occurrence and Term Frequency-Inverse Document Frequency (TF-IDF). The experiments showed that linear SVM performed better than Naïve Bayes [107].

In 2014, Yang and Cardie suggested a context-aware method for analyzing sentiments at the sentence level. The dataset comprised of sentences labeled with positive, negative or neutral and unlabeled sentiments. The researchers used the concept of posterior regularization to develop a set of context-aware posterior constraints using lexical and discourse knowledge [108].

In 2015, Chikersal, Poria and Cambria performed sentiment analysis combining a rule-based classifier with SVM. The rule-based classifier classified emoticons and opinion words in tweets as positive, negative or unknown. SVM classified tweets as positive, negative and neutral. The two classifiers were combined to fine-tune the predictions of the SVM classifier. The output showed that the rule-based classifier refined the predictions of SVM [109].

Event evaluation and monitoring are critical to event planning. These two work together to determine the success of an event. Event evaluation is the process of measuring the success of an event whereas monitoring involves collecting data about the event. There are two main approaches to event evaluation. The *quantitative* approach is factual and mostly based on figures such as the attendance levels and the profit generated. Questionnaires and surveys collect data for this approach. A questionnaire provides the opportunity to reach a large number of people compared to interviews via emails or on site and it is cheaper than interviews as it is self sufficient. *Qualitative* approach tends to be general as the population selected is varied and large though there may be low response rate. It does not explain complex issues or interactions [110]. The quantitative approach generates data for numerical analysis revealing patterns in the collected data. The qualitative approach captures the opinion and attitude of people through interviews, observations, focus groups etc. Interviews provide a medium to interact directly and openly with stakeholders of the event. *Interviews* are a valuable source of information that cannot be gathered from surveys as interviewers are able to express themselves. However, they are time consuming and expensive. *Observations* help to check the gestural expressions of participants, comprehend communication among participants and check the time spent on activities [111, 112]. They are also used to gather data about the attendees' participation. Observations collect data on how people partake in activities, however they do not explain their behavior. Observations may not yield accurate results as observers tend to be biased and people may not be genuine and behave differently when they know they are watched [113]. A *focus group* is an informal discussion among a group of 6 to 8 people with similar concerns, experiences or backgrounds concerning a particular topic [114]. Focus groups seek to explore and probe the thoughts of people, their thinking processes and patterns. A qualitative approach provides contextual data that aids in understanding complex issues and interactions. It aims to understand a stakeholder's perspective. Though this approach is time consuming, it generates rich and detailed data for deeper understanding of the opinions of people about the event. [115]. This approach may generate biased data.

Analysing event data uses quantitative and qualitative data analysis methods. *Quantitative data analysis* includes but is not limited to measuring central tendencies by dispersion, correlation, cross-tabulation, and frequency tables. Modes, median and mean are measures of central distribution. Dispersion measures the variations that exist in data by calculating the standard deviation, range, and inter-quartile range. Correlation measures the strength of the relationship existing between two variables. It assesses how an increase or decrease in a variable affects another variable. Frequency tables visually display the summary as the count of observations in tables. *Qualitative data analysis* identifies, analyses and interprets the uncovered patterns in data. Qualitative data analysis includes content analysis, thematic coding, and framework matrices. *Content analysis* is a technique used to compress the words in a text into content categories based on some rules of coding and can be used to describe the focus of individuals and groups of people [116]. Content analysis compresses the opinions into content categories based on evaluation questions. Thematic coding identifies texts or images that share a common link or idea and indexes them into categories to form a thematic framework [117]. *Framework matrices* summarize and synthesize text into cases and themes. Cases correspond to rows and themes correspond to columns in the matrix that the framework employs. It sorts data according to cases by the identified themes in the text [118].

There are a number of factors to consider when planning events. These include allocating budget, selecting venue, getting permits, coordinating transport, parking and security, arranging for speakers or entertainers among others. A blend of these factors is necessary to ensure the success of the event. According to [119], the memory of the experience of an event is the hallmark of the event because experiences leave a lasting impression on attendees, clients, among other stakeholders. Delivering an experience to clients is not attained only by serving the client but also by building a relationship with them and engaging clients on individual level. Customer satisfaction is equally important. Organizations that pay critical attention to and manage customer experience successfully improve customer satisfaction [120]. Currently, social media stands out as one of the most vibrant sources of opinions on products and services.

Social media and social networks are embracing the cultural events industry. Research conducted by Pew Research Center reported that social media had a tremendous impact on marketing strategies, educational efforts and overall performance [120].

Using qualitative and quantitative data analysis methods to evaluate the event provides valuable information from the opinions expressed; however, it does not factor in the possibility of the existence of sentiments in the expressed opinions. Our focus is collecting social network data to further evaluate the cultural events in an agile way.

Social networks provide data to evaluate the success of an event via the opinions expressed by attendees as a form of feedback. Sentiment analysis of the feedback involves more than just analyzing the number of likes, shares and comments from social networks. It deals with client feelings rather than post contents [121].

The proposed architecture for cultural event evaluation uses data received from social networks. Prior to sentiment analysis, the noisy and unstructured data is cleaned. This process is multi-level; it includes document, sentence, word, and aspect levels. The results are meaningful for decision-making and agility improvement.

This is because each of the above four levels helps to evaluate the event in certain detail. It provides knowledge to improve future events and increases agility. For example, a comment: "Love to see all the photos working together. Thanks for shooting" represents a tweet. We treat a tweet as one sentence. This tweet contains a sentiment. Intuitively, we know the sentiment in this tweet is positive because it has a positive sentiment word "love" which we can identify at the word level. This is an easy case; however, certain cases are harder to classify. The positive sentiment is about photography identified at the aspect level. This aspect level helps to locate the exact aspect that an attendee liked/disliked. The document level provides a general overview of the overall sentiment in a set of tweets. This level does not provide details about the tweets classified as negative, positive or neutral; instead, it provides an overall view. The four levels will provide knowledge for different kinds of stakeholders to satisfy their specific requirements.

We collect the data from social networks through the custom APIs. These APIs get the data from the social networks easier; however, there are limitations to the data size. In our case, the data deal with the Twitter Search API on the Beale Street Music festival. Beale festival is one of the three events celebrated in Memphis, Tennessee. It has been celebrated annually since 1977. The history of Beale Street Music dates back to 1800 when Afro-American musicians from the South came to Beale Street and performed there [122].

The data from social networks needs to be cleaned and prepared before we analyze it. Choosing between machine learning approach and lexicon-based approach is a challenge. The supervised machine learning approach requires pre-classified data (i.e. training data) to classify the unclassified data (i.e. new dataset); however, currently there is no open dataset on cultural festivals. Also, a semantically different dataset from another problem domain cannot be used as training data. One option is to manually classify the data; this is often time consuming and not enough accurate. The lexicon-based approach requires resources; however, it does not require a pre-classified data set. The existing lexical resources may not be applicable as their terms may be different from our problem domain. We decided to use the machine learning-based Sentiment140 API to classify sentences by their polarities as it successfully classifies data from different domains.

Beale Street Music festival was evaluated at the sentence level using R's Sentiment140, one of the R packages that can be used for sentiment analysis. This package used the API that employed Maximum entropy to classify tweets. Cleaned tweets were passed to the Sentiment function. The function, in turn, returned the tweets together with their polarities (i.e. negative, positive and neutral). Each tweet representing a sentence was classified by the sentiment function as negative, positive or neutral. A total of 5,000 tweets were collected for the event using the twitter search API. Out of these tweets, 376 were classified as negative, 3985 were classified as neutral and 639 were classified as positive. Nearly 80% of sentiments were classified correctly.

We performed sentiment analysis at the document level using the polarities of tweets as in [135]. In our case, document level aggregated the polarities of each tweet. Then, we used the following ratio to compare the polarities in the document: the ratio of positive tweets to negative tweets to neutral tweets. From our sentiment classification at the sentence level, we know that 639 tweets were classified as positive, 3985 were classified as neutral and 376 tweets were classified as negative resulting in a simplified 80:47:498 ratio out of every 625 tweets. As shown in Figs. 1.2 and 1.3, the majority of the tweets collected were classified as neutral. Analyzing these tweets revealed that they mainly contained information about the event such as weather updates, dress code information for the event and tips on getting ready for the rain, because the rain delayed the start of the event. Interestingly, there was a sharp decline in the number of tweets of May 2–3 for all the polarity categories. It is clear that there were more positive tweets for each day than negative tweets which indicated that the visitors generally enjoyed the festival.

At the word level, we extracted the positive, negative and neutral words using the positive and negative words from Hu and Liu's opinion dictionary. Figure 1.4

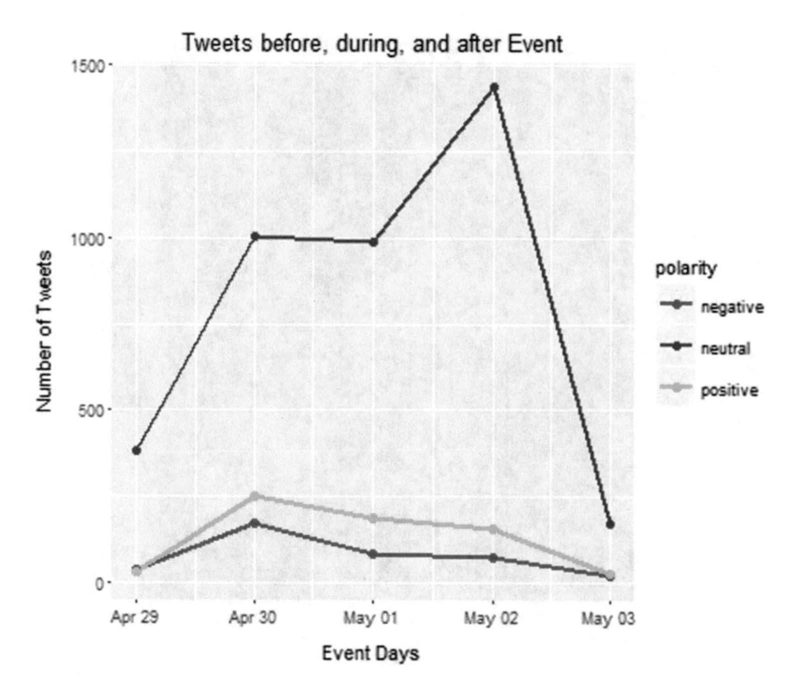

Fig. 1.2 Tweets, polarity and time

is a comparison word cloud that displays the words used in the tweets displayed categorically. Our word cloud depicts a set of words taken from the tweets and is grouped based on the polarity of the individual words used within the tweets. However, using just the individual words to analyze the sentiments shared by people turns not to be very efficient. A tweet such as "I am not happy with this festival" contains within it one negative word "not", one positive word "happy" and the other neutral words. These words in isolation do not provide any meaningful information for decision-making by event organizers. On the contrary, however, the entire text provides a sentiment that is negative. We agree with [98] that the word level analysis of text is insufficient for a positive contribution to opinion mining. To further improve we recommend adding higher levels of sentiment analysis.

We are going to further enhance the analysis by ML algorithms. These include: Naïve Bayes, Decision Trees, Random forests, Neural Networks, Boosting and Bagging algorithms.

We demonstrated how sentiment analysis can be used for evaluating cultural events. We proposed a sentiment architecture that demonstrates how sentiment analysis can be performed at different levels to provide details for different information needs. We proceeded to perform sentiment analysis at four different levels. Though not all tweets at the sentence level were classified correctly, our application provided the initial output for further drilling down at the aspect level as well as an overall

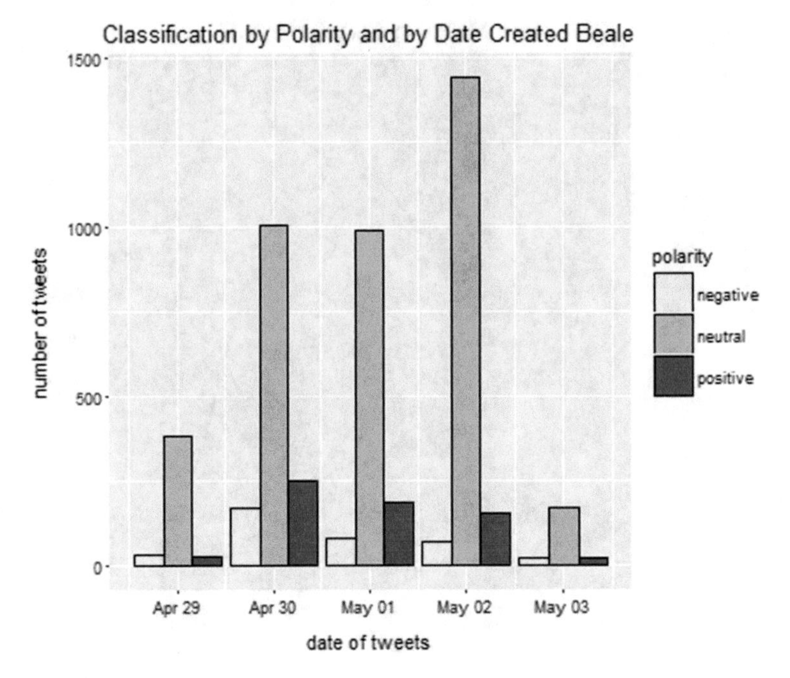

Fig. 1.3 Polarity of Tweets per day

Fig. 1.4 Comparison cloud of positive, negative and neutral words in Tweets

impression at a higher level without excessive detail. The word level gives an idea of the kind of sentiment words used by participants.

We used sentiment analysis to evaluate events qualitatively using data collected from Twitter. As the aim of qualitative evaluation is to understand the opinions of participants, sentiment analysis helps us to understand the participant's experience and opinions, and to use their opinions for improving the agility of cultural events in the future. The proposed multi-level architecture is easily adjustable (i.e. agile) to suit the needs of event planning organizations. In future, we plan to develop an ML-based algorithm for sentiment analysis to suit event evaluation using sentiment analysis more accurately.

1.6 Conclusion: How Agile Way Works

This chapter introduced the phenomenon of crisis in software development, which basically resulted from a sharp misbalance between project resources and business requirements/constraints. This gave us an initial understanding of this essential concept and its vital role in large-scale software development, as it often becomes mission-critical in enterprises. To cope with the crisis identified, we suggested the agility framework. Within this framework, we detected such core dimensions as technological, business and individual agility, and traced their links to smart enterprise management. To illustrate the basic concepts of crisis and agility, we discussed the possible applications and evolution of this agile framework in large-scale complex systems (such as bridges), and presented a few important examples of this agile way in Russia (for additional reading on the other environments such as Europe, and the USA, see [32, 123]). We also overviewed the Industry 4.0 smart digital transformation as a way to improve enterprise business agility and supported that by the case studies which included decentralized applications in general and blockchain in particular. Afterwards, we demonstrated the application of crisis-responsive feedback methods based on human factor-related sentiment analysis and machine learning.

The above mentioned ideas and case study outcomes provide a general framework for smart, crisis-responsive management in enterprise-scale systems. However, at this point our sketch of this big picture looks brief and incomplete. Therefore, in the next chapters, we will concentrate on agility in software development and provide more helpful examples of crisis management in developing resistant and responsive software solutions. To do this, we will consider agile processes, patterns, practices and case studies, including a deeper investigation of the technological and human factors that often appear to be the root cause of a crisis. Taken together, our findings will form a handy toolkit for a crisis manager of smart software development.

References

1. Tech Trends 2016: Innovating in the digital era (2016). Deloitte University Press. https://www2.deloitte.com/content/dam/Deloitte/lu/Documents/technology/lu_en_techtrends_lux_15042016.pdf.
2. http://www.rbc.ru/photoreport/22/04/2016/57189af89a79474349d162b9.
3. Nagel, R. (1991). *21st century manufacturing enterprise strategy: An industry-led view*. Diane Pub Co.
4. Sambamurthy, V., Bharadwaj, A., & Grover, V. (2003). Shaping agility through digital options: Reconceptualizing the role of information technology in contemporary firms, *MIS Quarterly*, *27*(2).
5. Lu, Y., & Ramamurthy, K. (2011) Understanding the link between information technology capability and organizational agility: An empirical examination. *Management Information Systems Quarterly*, 931–954.
6. Mathiassen, L., & Pries-Heje, J. (2006). *European Journal of Information Systems, 15,* 116. https://doi.org/10.1057/palgrave.ejis.3000610.
7. van Oosterhout, J., Heugens, P. P. M. A. R., & Kaptein, S. P. (2006). The internal morality of contracting: advancing the contractualist endeavor in business ethics. *Academy of Management Review, 31,* 521–539.
8. Overby, et al. (2006). Enterprise agility and the enabling role of information technology. *European Journal of Information Systems, 15*(2), 120–131. https://doi.org/10.1057/palgrave.ejis.3000600.
9. Kumar, R. L. & Stylianou, A. C. (2014). A process model for analyzing and managing flexibility in information systems. *European Journal of Information Systems*, *23*(2), 151–184. Available at SSRN: https://ssrn.com/abstract=2405401 or http://dx.doi.org/10.1057/ejis.2012.53.
10. Chakravarty et al. (2013). *Supply chain transformation: Evolving with emerging business paradigms* (2,014th ed.). Springer Texts in Business and Economics, Kindle Edition.
11. Lowry, P. B. & Wilson, D. (2016). Creating agile organizations through IT: The influence of internal IT service perceptions on IT service quality and IT agility. *Journal of Strategic Information Systems (JSIS)*, *25*(3), 211–226. Available at SSRN: https://ssrn.com/abstract=2786236.
12. Chen, C., Liao, J., & Wen, P. (2014). Why does formal mentoring matter? The mediating role of psychological safety and the moderating role of power distance orientation in the Chinese context. *International Journal of Human Resource Management, 25*(8), 1112–1130.
13. Richardson, S. M., Kettinger, W. J., Banks, M. S., & Quintana, Y. (2014). IT and agility in the social enterprise: A case study of st jude children's research hospital's "Cure4Kids" IT-platform for international outreach. *Journal of the Association for Information Systems*, *15*(1), Article 2.
14. de Albuquerque, P., & Christ, M. (2015). The tension between business process modeling and flexibility, *The Journal of Strategic Information Systems*, *24*(3), 189–202, https://doi.org/10.1016/j.jsis.2015.08.003.
15. Kärcher, B. (2015). Alternative wege in die industrie 4.0–möglichkeiten und grenzen. In A. Botthof, & E. A. Hartmann (Eds.), *Zukunft der Arbeit in Industrie 4.0* (pp. 47–58). Berlin: Springer.
16. Lay, G. (Ed.). (2014). *Servitization in industry*. Switzerland: Springer International Publishing.
17. Lüdtke, A. (2015). Wege aus der Ironie in richtung ernsthafter automatisierung. In A. Botthof, & E. A. Hartmann (Eds.), *Zukunft der Arbeit in Industrie 4.0* (pp 125–146). Berlin: Springer.
18. Rigby, D., & Bilodeau, B. (2015). *Management tools & trends 2015*. London: Bain and Company.
19. Roos, G. (2015). Servitization as innovation in manufacturing—a review of the literature. In *The handbook of service innovation* (pp. 403–435). London: Springer.
20. Cline A. (2015). *Agile development in the real world*. New York: Apress, 297 s. ISBN 978-1-4842-1678-1.

21. Schmidt, T. S., & Paetzold, K. (2017). Challenges of agile development: A cause-and-effect analysis. In G. Fanmuy, E. Goubault, D. Krob, & F. Stephan (Eds.), *Complex systems design & management*. Cham: Springer.
22. Dingsøyr, T., & Moe, N. B. (2014). Towards principles of large-scale agile development. In: *Agile Methods. Large-Scale Development, Refactoring, Testing, and Estimation. XP 2014 International Workshops*, Rome, Italy, May 26–30, 2014, Revised Selected Papers.
23. Dyer, L., & Ericksen, J. (2009). Complexity-based agile enterprises: Putting self-organizing emergence to work. In A. Wilkinson et al. (Eds.), *The sage handbook of human resource management* (pp. 436–457). London: Sage.
24. Gromoff, A., Bilinkis (Stavenko), J., Shumsky, L. (2016). Vision of multilevel modelling of processes in enterprise architectures affected by big data collection and analysis. In *International Conference Information Systems 2016 Special Interest Group on Big Data Proceedings*. 10.
25. Versteeg, G., & Bouwman, H. (2006). *Information Systems Frontiers, 8,* 91. https://doi.org/10.1007/s10796-006-7973-z.
26. IEEE 1471. ANSI/IEEE 1471-2000, Recommended practice for architecture description of software-intensive systems.
27. ISO 15704:2000. Industrial automation systems—Requirements for enterprise-reference architectures and methodologies.
28. Zachman. (1987). A framework for information systems architecture. *IBM Systems Journal, 26*(3), p. 276.
29. TOGAF ADM. http://www.opengroup.org/subjectareas/enterprise/togaf.
30. Hoogervorst, J. (2002). HR strategy for ICT-driven business context. *International Journal of Human resources management*. *13*(8), pp. 1245–1265.
31. Gromoff, A. (2010). A study of the subject-oriented approach for automation and process modeling of a service company. *S-BPM ONE, 2010,* 192–206.
32. Naur, P., & Randell, B. (1968). Software engineering. In *Report on a conference sponsored by the NATO science committee*. Garmisch, Germany, 7–11 Oct 1968.
33. http://www.rbc.ru/research/society/24/05/2016/573de5139a79478774746561.
34. Chaum, D. (1981). Untraceable electronic mail, return addresses, and digital pseudonyms. *Communications of the ACM, 24*(2), 84–90.
35. Clarkson, M. R., Chong, S., & Myers, A. C. (2008). Civitas: Toward a secure voting system. In *IEEE Symposium on Security and Privacy* (pp. 354–368).
36. Hao, F., Kreeger, M. N., Randell, B., Clarke, D., Shahandashti, S. F., & Lee, P. H. -J. (2014). Every vote counts: Ensuring integrity in large-scale electronic voting. *The USENIX Journal of Election Technology and Systems, 1*.
37. Kiayias, A., Korman, M., & Walluck, D. (2006). An internet voting system supporting user privacy. In *Computer Security Applications Conference, 2006. ACSAC'06. 22nd Annual* (pp. 165–174).
38. Adida, B., Marneffe, O., Pereira, O., & Quisquater, J. -J. (2009). Electing a university president using open-audit voting: Analysis of real-world use of helios. In *Proceedings of the Electronic Voting Technology Workshop/Workshop on Trustworthy Elections (EVT/WOTE)*. USENIX Association.
39. Andrew C. (2004). Practical high certainty intent verification for encrypted votes.
40. Ryan, P. Y. A., Bismark, D., Heather, J., Schneider, S., & Xia, Z. (2009). Prêt à voter: a voter-verifiable voting system. *IEEE Transactions on Information Forensics and Security, 4*(4), 662–673.
41. Shahandashti S. F., & Hao F. (2016). DRE-ip: A verifiable evoting scheme without tallying authorities. In *European Symposium on Research in Computer Security* (pp. 223–240). Springer.
42. Chaum, D. (2004). Secret-ballot receipts: True voter-verifiable elections. *IEEE Security & Privacy, 2*(1), 38–47.
43. Bell, S., Benaloh, J., Byrne, M. D., DeBeauvoir, D., Eakin, B., Fisher, G., et al. (2013). STAR-Vote: A secure, transparent, auditable, and reliable voting system. *USENIX Journal of Election Technology and Systems, 1*(1), 18–37.

44. Chaum, D., Carback, R., Clark, J., Essex, A., Popoveniuc, S., Rivest, R. L., et al. (2008). Scantegrity II: End-to-End verifiability for optical scan election systems using invisible ink confirmation codes. *EVT, 8,* 1–13.
45. Groth, J. (2004). Efficient maximal privacy in boardroom voting and anonymous broadcast. In *International Conference on Financial Cryptography* (pp. 90–104). Springer.
46. Kiayias, A., & Yung, M. (2002). Self-tallying elections and perfect ballot secrecy. In *International Workshop on Public Key Cryptography* (pp. 141–158). Springer.
47. Cybersecurity case study competition (2016). http://www.economist.com/whichmba/mba-case-studies/cybersecurity-case-study-competition-2016. Accessed 07 Nov 2017.
48. McCorry, P., Shahandashti, S. F., & Feng, H. A smart contract for boardroom voting with maximum voter privacy.
49. Feng, H., Ryan, P. Y., & Zielinski, P. (2010). Anonymous voting by two-round public discussion. *IET Information Security*, *4*(2), 62–67.
50. Kosba, A., Miller, A., Shi, E., Wen, Z., & Papamanthou, C. Hawk: The blockchain model of cryptography and privacy-preserving smart contracts.
51. Yang, N., & Clark, J. Practical governmental voting with unconditional integrity and privacy.
52. Kohno T., Stubblefield A., Rubin A. D., & Wallach D. S. (2004). Analysis of an electronic voting system. In *IEEE Symposium on Security and Privacy* (pp. 27–42).
53. Tatsuaki O. (1998) Receipt-free electronic voting schemes for large scale elections. In: *Security Protocols, 5th International Workshop* (Vol. 1361, pp. 25–35).
54. Rivest, R. L., Shamir, A., & Adleman, L. (1978). A method for obtaining digital signatures and public-key cryptosystems. *Communications of the ACM, 21*(2), 120–126.
55. Nazmul, I., et al. (2017). A new e-voting scheme based on revised simplified verifiable re-encryption mixnet. In *Paper presenter in Networking, Systems and Security (NSysS), 2017 International IEEE Conference.*
56. Meenal, J., & Singh, M. (2017) Identity based secure RSA encryption system. In: *Proceedings of International Conference on Communication and Networks*. Springer: Singapore.
57. Chugunkov, I. V. et al. (2017). Classification of pseudo-random number generators applied to information security. In: *Proceedings in Young Researchers in Electrical and Electronic Engineering (EIConRus), 2017 IEEE Conference of Russian.*
58. Olivier, P., & Ronald, L. Rivest. Marked Mix-Nets.
59. King-Hang W., et al. A review of contemporary E-voting: Requirements, technology, systems and usability.
60. Schneider A., Meter C., & Hagemeister P. (2017) Survey on remote electronic voting. arXiv preprint arXiv:1702.02798.
61. Ali, T. S., & Murray J. (2016) An overview of end-to-end verifiable voting systems. In *Real-world electronic voting: Design, analysis and deployment* (pp. 171–218). CRC Press.
62. Herschberg, Mark A. (1997). *Secure electronic voting over the world wide Web*. Massachusetts Institute of Technology: Diss.
63. Neumann, S., Budurushi, J., & Volkamer, M. (2014). Analysis of security and cryptographic approaches to provide secret and verifiable electronic voting. In *Design, development, and use of secure electronic voting systems*. IGI Global.
64. Jonker, H., Mauw, S., & Pang, J. (2013). Privacy and verifiability in voting systems: Methods, developments and trends. *Computer Science Review, 10,* 1–30.
65. Goldwasser, S., Micali, S., & Rackoff, C. (1989). The knowledge complexity of interactive proof systems. *SIAM Journal on Computing, 18*(1), 186–208.
66. Ewanick, B. (2011). Zero knowledge proof.
67. Bootle J., et al. (2016). Efficient zero-knowledge proof systems. In *Foundations of security analysis and design VIII* (pp. 1–31). Springer International Publishing.
68. Meter, C. (2017). Design of distributed voting systems. arXiv preprint arXiv:1702.02566.
69. Capitanio, I. (2016). How technology can transform the sports domain. http://www.vocaleurope.eu/how-technology-can-transform-the-sports-domain-blockchain-2-0-will-boost-anti-doping-fight-sports-investments-and-e-sports/. Accessed 07 Nov 2017.

70. Nakamoto, S. (2008) Bitcoin: A peer-to-peer electronic cash system. https://Bitcoin.org/Bitcoin.pdf. Accessed 06 Nov 2017.
71. Trends: Bitcoin Continues to Struggle, Ethereum Looks Strong (2017). https://cointelegraph.com/news/trends-Bitcoin-continues-to-struggle-Ethereum-looks-strong. Accessed 06 Nov 2017.
72. Levich, A. (2015). Microsoft azure: The blockchain sandbox. http://www.coinfox.info/news/reviews/5231-microsoft-azure-the-blockchain-sandbox. Accessed 07 Nov 2017.
73. Szabo, N. (1997). The idea of smart contracts. http://www.fon.hum.uva.nl/rob/Courses/InformationInSpeech/CDROM/Literature/LOTwinterschool2006/szabo.best.vwh.net/idea.html. Accessed 07 Nov 2017.
74. Luu, L., Chu, D. -C., Olickel, H., Saxena, P., & Hobor, A. Making smart contracts smarter. https://www0.comp.nus.edu.sg/~hobor/Publications/2016/Making%20Smart%20Contracts%20Smarter.pdf. Accessed 07 Nov 2017.
75. Dhaliwal S. (2017) Ethereum blockchain will help un go cashless, distribute food to hungry in Jordan. https://cointelegraph.com/news/Ethereum-blockchain-will-help-un-go-cashless-distribute-food-to-hungry-in-jordan. Accessed 07 Nov 2017.
76. Liu, B. (2012). *Sentiment analysis and opinion mining*. Morgan & Claypool Publishers.
77. Pang, B., & Lee, L. (2008). *Opinion mining and sentiment analysis*.
78. Kaplan, A. M., & Haenlein, M. (2012). Users of the world, unite! The challenges and opportunities of social media. *Business Horizons, 53*(1), 59–68.
79. Edwards, J. (2016). Leaked Twitter API data shows the number of tweets is in serious decline. Retrieved November 7, 2017, from Business Insider: http://www.uk.businessinsidercom/tweets-on-twitter-is-in-serious-decline-2016-2.
80. Culture, C. (2013). The importance of cultural heritage. Retrieved November 7, 2017, from Cultivating Culture: http://www.cultivatingculture.com/2013/04/05/the-importance-of-cultural-heritage/.
81. Blog, M. R. (2013). Technology helps preserve biodiversity and traditiional cultures. Retrieved November 7, 2017, from Microsoft: https://blogs.msdn.microsoft.com/msr_er/2013/06/28/technology-helps-preserve-biodiversity-and-traditional-cultures/.
82. Hugo, N. C., & Lacher, R. G. (2014). Understanding the role of culture and heritage in community festivals: An importance-performance analysis. *Journal of Extension*, *52*.
83. Gertz, D. (1991). *Festivals, special events, and tourism*.
84. Rice, D. R., & Zorn, C. (2013). Corpus-based dictionaries for sentiment analysis of specialized vocabularies. In *New Directions in Analyzing Text as Data Workshop*. London.
85. Wang, S., & Manning, C. D. (2012). Baselines and bigrams: simple, good sentiment and topic classification. In *Proceedings of the 50th Annual Meeting of the Association for Computational Linguistics*.
86. Gamallo, P., & Garcia, M. (2014). Citius: A naive-bayes strategy for sentiment analysis on english tweets. In *Proceedings of the 8th International Workshop on Semantic Evaluation*, (pp. 171–175). Dublin, Ireland.
87. Mehra, N., Khandelwal, S., & Patel, P. (2012). *Sentiment identification using maximum entropy analysis of movie reviews*. USA.
88. Lehrer, A. (1985). Markedness and antonymy. *Journal of Linguistics*, 397–429.
89. Turney, P. (2002). Thumbs up or thumbs down? Semantic orientation applied to unsupervised classification of reviews. In *ACL 2002* (pp. 417–424).
90. Baccianella, E. A., Sebastiani, S., & Sebastiani, F. (2010). An enhanced lexical resource for sentiment analysis and opinion mining. *LREC, 10,* 2200–2204.
91. Wiebe, J., & Cardie, C. (2005). Annotating expressions of opinions and emotions in language. *Language Resources and Evaluation*, 165–210.
92. Strapparava, C., & Valitutti, A. (2004). Wordnet affect: An affective extension of wordnet. *LREC, 4,* 1083–1086.
93. Moreno-Ortiz, A., & Hernández, C. P. (2013). Lexicon-based sentiment analysis of twitter messages in Spanish. Procesamiento del Lenguaje Natural, Revista, 93–100.

94. Musto, C., Semeraro, S., & Polignano, M. (2014). A comparison of lexicon based approaches for sentiment analysis of microblog posts. Information filtering and retrieval. In *Proceedings of the 8th International Workshop on Information Filtering* (pp. 12–23).
95. Momtazi, S. (2012). *Natural language processing sentiment analysis*.
96. Bonev, B., Sanchez, R., & Rojas, S. O. (2012). Opinum: statistical sentiment analysis for opinion classification. In Computational approaches to subjectivity and sentiment analysis, (pp. 29–37). Jeju.
97. Jurafsky, D., & Martin, J. (2014). *Speech and language processing*.
98. Cambria, E., Schuller, B., Xia, Y., & Havasi, C. (2013). *Knowledge-based approaches to concept- level sentiment analysis*. IEEE Computer Society.
99. Bing, L. (2015). *Opinions, sentiment, and emotion in text*. Cambridge University Press.
100. Gundla, V. A., & Otari, M. S. (2013). A review on sentiment analysis and visualization of customer reviews. *International Journal of Science and Research, 4*.
101. Chunping, O., Yongbin, L. Shuqing, Z., & Yang Xiaohua, Y. (2015). Features-level sentiment analysis of movie reviews. *Advanced Science and Technology Letters*, 110–113.
102. Jagtap, V. S., & Pawar, K. (2013). Analysis of different approaches to sentence-level sentiment. *International Journal of Scientific Engineering and Technology*, 164–170.
103. Steinberger, J., Brychcin, T., & Konkol, M. (2014). Aspect-level sentiment analysis in Czech. In *Computational Approaches to Subjectivity, Sentiment and Social Media Analysis*. Baltimore.
104. Bongirwar, V. (2015). A survey on sentence level sentiment analysis. *International Journal of Computer Science Trends and Technology (IJCST), 3*(3).
105. Zhang, L., & Liu B. (2014). Aspect and entity extraction for opinion mining. In *Data mining and knowledge discovery for big data*. Springer.
106. Tan, C., Lee, L., Tang, J., Jiang, L., Zhou, M., & Li, P. (2011). *User-level sentiment analysis incorporating social networks*. San Diego, California: KKD.
107. Tripati, G., & Naganna, S. (2015). Feature selection and classification approach for sentiment analysis. *Machine Learning and Applications: An International Journal, 2*.
108. Yang, B., & Cardie, C. (2014) Context-aware learning for sentence-level sentiment analysis with posterior regularization. https://www.aclweb.org/anthology/P/P14/P14-1031.xhtml. Accessed Nov 6 2017.
109. Chikersal, P., Poria, S., & Cambria, E. (2015). SeNTU: Sentiment analysis of tweets by combining a rule-based classifier. http://anthology.aclweb.org/S/S15/S15-2108.pdf. Accessed Nov 6, 2017.
110. Garbarino, S., & Holland, J. (2009). In J. Holland (Ed.) *Quantitative and qualitative methods in impact evaluation and measuring results*. Social Development Dialect.
111. Schmuck, R. (1997). *Practical action research for change*. IRI/Skylight Training and Publishing.
112. An evaluation toolkit for e-libary developments: http://www.evalued.bcu.ac.uk/tutorial/4c.htm.
113. Prevention, C. F. (2008). *Evaluation briefs*.
114. Wilkinson, S. (2004). Focus groups: A feminist method. In S. Hesse-Biber (Eds.), *Femininst perspective on social research* (pp. 271–295). New York: Oxford University Press.
115. Mmari, K. (2006). *Using qualitative methods for monitoring and evaluation*. Johns Hopkins University.
116. Weber, R. (1990). *Basic content analysis*. California.
117. Mountain, A. (2014). Thematic coding. Retrieved November 7, 2017, from BetterEvaluation: http://betterevaluation.org/evaluation-options/thematiccoding.
118. Macfarlan, A. (2014). Framework matrices. Retrieved November 7, 2017, from BetterEvaluation: http://betterevaluation.org/evaluation-options/framework_matrices.
119. Pine, B. J., II, & Gilmore, J. H. (1999). *The experience economy*. Boston: Harvard Business School Press.
120. Duncan, E., Jones, C., & Rawson, A. (2013). The truth about customer experience. Harvard Business School Publishing Corporation. Retrieved from www.mckinsey.com.

121. Clarabridge. (2016). Retrieved November 7, 2017, from Sentiment Analysis Customer Experience Dictionary: http://www.clarabridge.com/sentiment-analysis/.
122. Brief History of Memphis in May (n.d.). Retrieved November 7, 2017, from Memphis in May International Festival: http://www.memphisinmay.org/about/history/.
123. Zykov, S. (2017). Crisis response and management: technological and human factor-based agility (Chapter 120/705). In M. Khosrow-Pour (Ed.), *Encyclopedia of information science and technology* (4th Ed., pp. 1396–1406). Information resources management association, USA, Vol. 2, Section: Crisis Response and Management, Release Date: July 2017, ©2018, 7500 pp.

Chapter 2
Agile Languages

Abstract This chapter classifies programming languages and outlines their application for software development. It overviews the domain specific languages and illustrates this concept with the case studies on inter-process communication and embedded system development. The focus is on further agility improvement by means of special purpose languages and crisis-responsive development environments.

Keywords Functional · Logical · Object-oriented · Domain specific language
Embedded system

2.1 Introduction: Communication Agility

This chapter presents a description of the programming languages and their application to software development. We compare several types of general purpose languages such as procedural, functional, logical, object-oriented, and a few others (the survey is based on [1]). For each language class, we identify strong and weak sides in terms of agility. As such, the primary goal of this chapter is to determine the applicability of each language family to mission-critical software development, especially in a crisis.

The early languages often addressed certain hardware architectures only; therefore, many of the computer systems of the old days were incompatible with each other and could not communicate efficiently. However, the software crisis revealed this problem as different generations of software and hardware should have been able to communicate with each other. This problem was partially solved by high level languages (the earlier examples included Fortran and LISP) and the later virtual machine-based languages (such as Java and C#).

Research will conclude that there is no universal language solution equally applicable to any software product development in crisis. However, our outcomes will help to combine and adjust the languages so that the target software products are more agile and crisis resistant.

S. V. Zykov, *Managing Software Crisis: A Smart Way to Enterprise Agility*,
Smart Innovation, Systems and Technologies 92,
https://doi.org/10.1007/978-3-319-77917-1_2

Often, it is the problem domain, which determines specific language choice to develop an agile product. Thus, we arrive at the concept of *domain specific languages*, or DSLs, which are developed for a particular domain of software application. This chapter presents an overview of DSLs and a couple of examples to illustrate the concept, such as inter-process communication and embedded system development.

The language applicability significantly depends upon the development team expertise in the problem domain (i.e. business constraints) and language proficiency (i.e. technical requirements). These are the human-related factors that may essentially promote or inhibit agility of the resulting software product. We discuss these human factors and the respective agile patterns and practices in Chap. 4 and [2].

This chapter is organized as follows. Section 2.2 overviews the programming language types and gives their brief classification. Section 2.3 presents an outline for a DSL-based software solution for inter-process communication. Section 2.4 discusses designing a DSL for general-purpose embedded system development. The conclusion summarizes the results of the chapter.

While discussing the languages let us focus on their agility that makes software products crisis resistant.

2.2 Why Languages?

The first programming languages appeared relatively recently with various research specified in the 1920s, 1930s and even 1940s. Out task is not to identify the correct date when the earliest language appeared, but rather focus on laws of their development.

As one would expect, first programming languages, as well as first computers, were rather primitive and focused on numerical computations. There were only theoretical scientific computations at that time (basically mathematical and physical), and applied problems, in particular, in the military field.

The programs written in early programming languages represented linear sequences of elementary operations with registers in which the data was stored.

It is necessary to note that early programming languages were optimized for hardware architecture of a specific computer for which they were intended and thus they provided high efficiency of computations, since standardization was not established. A program, which ran quite efficiently on one computer, often could not be executed on another.

Thus, early programming languages essentially depended on what is presently known as computational environment and roughly corresponded to modern machine codes or assembly languages.

The next decade was marked by the advent of the so-called high level programming languages, in comparison with their predecessors mentioned earlier, which were called low level languages accordingly.

Thus, the distinction is in the increase of efficiency of developers' work due to abstracting from specific features of hardware. One instruction (or operator) of a high level language corresponded to a sequence of several low level instructions (or commands). Recognizing that the program, as a matter of fact, represented a set of instructions given to a computer, such an approach to programming received the name *imperative*.

One more feature of high level languages was the opportunity to reuse program blocks written earlier by means of their identification and subsequent reference, for example, by name. Such blocks were named functions or *procedures*, through which programming has become more uniform and, therefore, agile.

With the high level language advent, the problem of dependence of implementation from hardware essentially decreased. As a payment for this, there appeared specialized software tools, transforming instructions of the source languages into machine code, also known as *compilers*. Computing speed went down however, this was compensated by an essential gain in the speed of application development and program code unification.

It is worth noting that operators and keywords of modern programming languages were more intelligent than the earlier meaningless strings of low level code, this also increased developer labor productivity.

Naturally, studying modern programming languages required significant expense of time and effort, and efficiency of implementation on outdated hardware was reduced. However, these were temporary difficulties and many of the first high level languages were so successfully implemented, that they are still actively used at present.

One of such examples is Fortran, a language that implemented formula-based computation algorithms. Another example is APL language, later transformed into BPL and then into C. The basic principles of the latter remained unchanged for several decades and are still present in C# language.

In the 1960s there appeared a new approach to programming which is an alternative to the imperative, and which is known as *declarative*. The essence of the approach is that a program represents description of actions rather than a set of commands.

The approach is much easier formalized by means of mathematics. Declarative programs are easier for debugging and verification. A high degree of abstraction is also an advantage of the approach as this promotes agility.

At the initial stage of development it was difficult for declarative programming languages to compete with the imperative because of the difficulty to create an efficient compiler implementation. Programs were slower, however, they could solve more abstract problems with smaller labor expenses. SML language, the direct predecessor of F# used for enterprise applications, has been developed as a means of the proof of theorems.

One branch of declarative programming was the *functional* approach, which has appeared with LISP language development (the name is derived from words 'LISt Processing').

A program written in functional language can be interpreted as a function. This approach enables transparent modeling of the source code. Under this approach, any program represents a function, some parameters of which can be also considered as functions. Thus, code reuse is similar to a function call.

Certain functions have variable types. This property provides *polymorphic* functions for heterogeneous data processing, which promotes agility.

One more important advantage of functional programming language implementation is automated dynamic computer memory disposal. Thus, the programmer gets rid of the routine necessity to supervise the data, by means of the *garbage collector* function.

Due to *pattern matching*, such functional languages as ML and Haskell are even more efficient for symbolic processing. Functional programming languages have a non-linear program structure, and they are often relatively slow.

In the 1970s, declarative programming language branch gave birth to *logical* programming languages. According to the logical approach to programming, the program represents a set of rules or logical statements. Logical programming languages are based on classical logic and are applicable for systems of logic reasoning, particularly, for expert systems. They are applicable for decision-making rules, for example, in business-oriented applications, i.e. address the topmost level of the enterprise agility matrix, see Fig. 1.1).

The important advantage of the approach is *backtracking*, i.e. returning to previous sub-goals after a negative result of the analysis of one of the variants during a decision search (like in a game of chess). That eliminates the necessity of a brutal search for decision-making and increases efficiency and agility of implementation.

A drawback of the logical programming approach is inefficient real-time implementations e.g. for mission-critical systems.

Examples of logical programming languages include PROLOG (the name is derived from words ‘PROgramming in LOGic’) and Mercury.

In the *object-oriented* approach, the program describes objects, their properties (or attributes), sets (or classes), relations between them, ways of their interaction and operations with objects (or methods).

Inheritance of attributes and methods allows to build derivative concepts, and to create a problem domain model of virtually any complexity and properties.

A theoretically interesting and practically important property of the object-oriented approach is supporting an *event* mechanism that changes attributes of objects, and models their interaction in a problem domain.

Moving along the hierarchy of classes from more general concepts of a problem domain to more specific ones (or from more complex to more simple ones), the developer gets an opportunity to change an abstraction degree of the real world representation model.

Objects, classes and methods can be polymorphic, that makes the software more flexible and universal, and thus improves product agility.

Complexity of the formal object models results in difficulties of object-oriented software testing and verification.

The most known and widely used example of an object-oriented programming language is C++. Its direct descendant and logical continuation is C#.

In the 1990s, development of event-driven concepts of the object-oriented approach gave rise to scripting languages.

A program represents a set of data processing scripts, the choice of which is initiated by a certain event (such as mouse button click, cursor hit of a certain position, object attribute change, memory buffer overflow etc.). Events can be initiated by the operating system and its user.

The basic advantages and drawbacks of scripting languages are inherited from object-oriented programming languages.

The essential positive feature of a scripting language is its compatibility with advanced tools for design and implementation of the software, or Computer-Aided Software Engineering (CASE).

Languages that support parallel computation represent a set of descriptions of processes, which can be executed simultaneously or in a pseudo-parallel mode. In the latter case the CPU works in a time-sharing mode, allocating time for each of the concurrent processes as required.

Languages that support parallel computation improve agility of processing large data files and intensive data flows (e.g., high quality video). The other significant implementation scope of parallel computation languages are real-time systems where users must receive a system response directly after inquiry. Such systems are responsible for critical life-support and decision-making applications. Examples of the programming languages supporting parallel computation include Ada, Modula-2 and Oz.

The classification given is not absolute as programming languages emerge and evolve, and certain drawbacks are mitigated by advanced theoretical foundations and CASE tools.

Let us briefly enumerate the approaches to programming considered:

- Early non-structural approaches
- Structural or modular approach (the task is split into subtasks, then into algorithms, their block diagrams are made and implementation follows)
- Functional approach
- Logical approach
- Object-oriented approach

The new generation languages are typically more agile in terms of uniformity, handling versatile data and supporting concurrent processes; many of them focus on efficient enterprise-scale software development. The domain specific languages (DSLs) are used for solving the problem in a specific domain; they involve modeling (and so are referred to as DSML or domain specific modeling languages). According to [3], DSMLs are compact and convenient languages, which support the productivity of modeling and help to increase model quality and comprehensibility. In DSM, the solution is built on the idea of the problem domain by a developer or system analyst. It contains the syntax and semantics that model concepts at the same level of abstraction that the problem domain offers.

Below we are going to present a few domain specific cases which include development of new languages or extending currently existing ones.

2.3 Making Processes Communicate

This section describes the case study on improvement of inter-process communication technology also known as *Remote Procedure Call* (RPC). The concept of the proposed solution is based on the principle of intermediate interpreted layer. This layer aims to simplify the technology, unify the development process and improve code (re)usability.

This layer is used for the interface definition and communication process itself. The interface definition presents an object-oriented abstraction with methods of inter-process interaction. The interfaces are defined using a special DSL which provides a list of standard types for data transmission and a set of keywords for interface structure and data description. The DSL is extendable and it allows users to create complex data types using known types and define custom data types. The internal DSL is based on the interpreted language of the intermediate layer. This principle ensures simplification of the interface definition process by means of DSL and preserves the functionality of the base language. As an interpreter, the solution uses Lua language because of its efficient embedding property, straightforward syntax and high performance.

At the time of writing this book, there are certain issues common for RPC. These problems do not relate to the implementation but rather to basic architectural and technological principles.

The first issue is the invariability of interfaces. The connection between server and client parts of generated interfaces is very strong. Vendors of these RPC technologies do not provide means to change any end of the pair and save the connection, and they do not recommend doing that. It means that any changes of interface should be applied to both server and client. This drawback becomes a problem in case of several product versions with different interfaces. Each version of a client (or a server) should have the corresponding copy on the other side. Any change of interface leads to changes of communication protocol. If one side of a client-server system changes its interface the whole protocol becomes invalid because each side interprets the protocol in its own way.

The invariability of interfaces complicates product development and maintenance; it keeps RPC technologies out of enterprise software systems development. Another issue is that the communication protocol is fixed. This includes the type of protocol, the set of messages (and their semantics) and the principle of client-server message transmission. RPC technologies should support different environments and become agile.

CORBA technology uses General InterORB Protocol (GIOP) that provides implementations through TCP/IP(IIOP), SSL and HTTP [4]. Despite their diversity CORBA has a bulky overhead in terms of requests and overloads the network.

CORBA, ICE, GRPC and other protocols strictly determine how the function calls and other features should be presented in the network protocol. Each function call usually corresponds to one request. The request contains function and interface identifiers, information parameters and data. This principle does not take into account the nature of data or communication.

To optimize network consumption, the function should produce the request that first checks the state of server information instead of transmitting the data.

The user can select the protocol of communication only for the entire system or specific interface but not for the function call. However, agility requires combining the text and binary protocols.

Another issue is that communication logic is outside the communication layer. This problem increases the complexity of RPC usage. Any additional logic of communication is out of the middleware or RPC layer.

The ICE supports caching of function calls but this can be implemented only at the endpoint application layer. It means that the client is responsible for this.

The GRPC uses messages for communication and it represents the data as objects of the target interface. The users should implement interfaces of both client and server sides. These implementations should consider the protobuf (protocol buffer) message object creation and its lifetime. This allows for the fine tuning of communication requiring extra labor and deeper knowledge of the communication process, even for simple interfaces without any communication specifics (like caching).

Considering the runtime interface of RPC represents only the target language, there is only one way to implement product specific features: to share the features implementation as a module or library for target language. This approach is not suitable because of the complexity of the implementation.

RPC is complex enough even for relatively simple systems. For example, CORBA requires more than 200 lines of code for interface definition even though the same functionality can be provided in much less code [5]. These technologies provide native approaches to remote method call; however, they require a lot of extra logic to create and connect remote interfaces.

Problems detected are caused by architectural drawbacks. RPC technologies allow only to describe the semantics of the interface and static properties but not the logic of communication. The communication logic is encapsulated in the implementation.

Features are available only outside the communication layer, in the target application, the user takes all responsibilities to work with them.

The protocol is fixed, and the target system developers are unable to optimize the data transmission. The complexity of technologies is increased by the non-agile interface and redundant communication logic.

To make the interface more agile, we suggest a DSL [5].

The base language for the DSL, Lua, is a script language designed primarily for embedding. It is cross-platform and available for the major operating systems. Lua is a lightweight language with a simple syntax and a few data types. Basically, Lua is a prototype-oriented language; however, it can support other paradigms. It is often referred to as a *multi-paradigm* language [6].

Embedding is the strong side of Lua. This language is supplied with a binary interpreter and C library. The C API is powerful and easy to use. Lua has embedding support for a wide set of languages: C++, Java, C#, Python etc.

Lua has no multi-threading support. To execute several code blocks in parallel, Lua provides coroutines and threads, which are separate instances of execution stack and therefore cannot be executed in different system processes.

The way to efficiently work with Lua in multiple threads is to create several instances of Lua interpreter.

The Common Interface Definition (CID) layer aims to unify and simplify the technology. Logic of communication defined by user is placed in an interpretation layer and actually does not depend on the programming language of the target application. This approach decreases the cost of development and increases code reusability, i.e. improves agility.

Each component of the proposed system is a program block working as an independent job and presenting a specific interface for communication. Each component defines an integration scheme by itself and presents a list of supported integration schemes.

The following examples of the component manager's logic scheme assume that we already have a user component loaded, and it is going to load a 'messenger' component with a different integration scheme.

The first example represents in-state integration (see Fig. 2.1):

1. User makes a request to load a component messenger to API Manager
2. API Manager puts the data of the request to its own unique channel
3. API Manager makes a request to the host invoking the `system.cstorage` function and transmitting an ID of the channel
4. Host makes request to component manager for component loading with specific channel ID
5. Component manager reads the data from the specified channel, finds the required component package and places found package and its loader back into channel
6. Function returns to step 4
7. Function returns to step 3
8. API Manager reads the data returned from the channel
9. API Manager loads the raw component data and returns the loaded component (as object) to the user.

User makes a request to the loaded component to the messenger as it was a regular Lua table. The second example represents out-state integration (see Fig. 2.2):

1. User makes a request to load a component messenger to the API manager
2. API Manager puts the data of the request to its own unique channel
3. API Manager makes a request to the host invoking the `system.cstorage` function and transmitting an ID of the channel
4. Host makes a request to the component manager for component loading with a specific channel ID
5. Component manager reads the data from the specified channel and finds the required component package

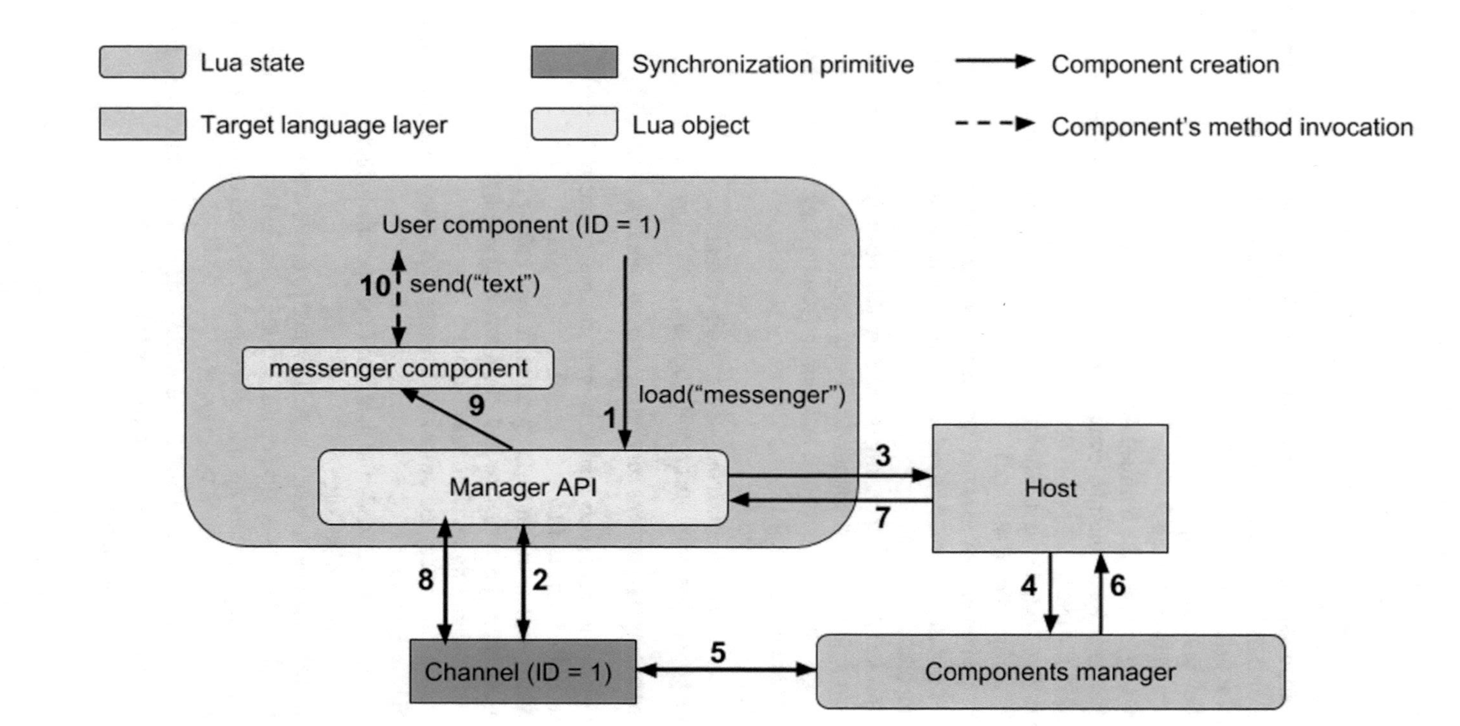

Fig. 2.1 In-state integration

6. Component manager makes a request to the host to create a new out-state component by calling `system.out-state_create` method transmitting the source code of required component
7. Host creates a new state and loads the received code
8. Function returns to step 6
9. The component manager puts component's connector back to the channel. Connector also includes a unique ID of channel to communicate with corresponding out-state component (channel with ID = 2)
10. Function returns to step 4
11. Function returns to step 3
12. API Manager reads the returned data from the channel
13. API Manager loads the raw component connector code and returns it (as an object) to the user
14. User makes a request to the loaded component messenger like to a regular Lua table (e.g. by a `send` method)
15. Component connector puts the input data into the channel
16. Component makes a request to the host to call a method for transmitting an ID of component instance
17. Host invokes the corresponding method for the messenger instance
18. Messenger instance reads input data from channel
19. Messenger handles the request and puts the results back into the channel
20. Function returns to step 17
21. Function returns to step 16
22. Messenger connector reads output data from channel
23. Messenger connector returns results to user.

The third example represents service integration (see Fig. 2.3):

1. User makes a request to load a component messenger to the API manager presented in the same state
2. API Manager puts the data of request to its own unique channel
3. API Manager makes a request to the host invoking `system.cstorage` function and transmitting an ID of the channel
4. Host makes a request to component manager for component loading with a specific channel ID
5. Component manager reads the data from the specified channel and finds the required component package
6. Component manager makes a request to host to create a new out-state component by calling `system.service_create` method transmitting the source code of the component loader and component. The component loader code also includes information about the channel
7. Host creates a new state, loads the received code to it and runs a new thread in which the `entry` method of the component is invoked
8. Function returns to step 6

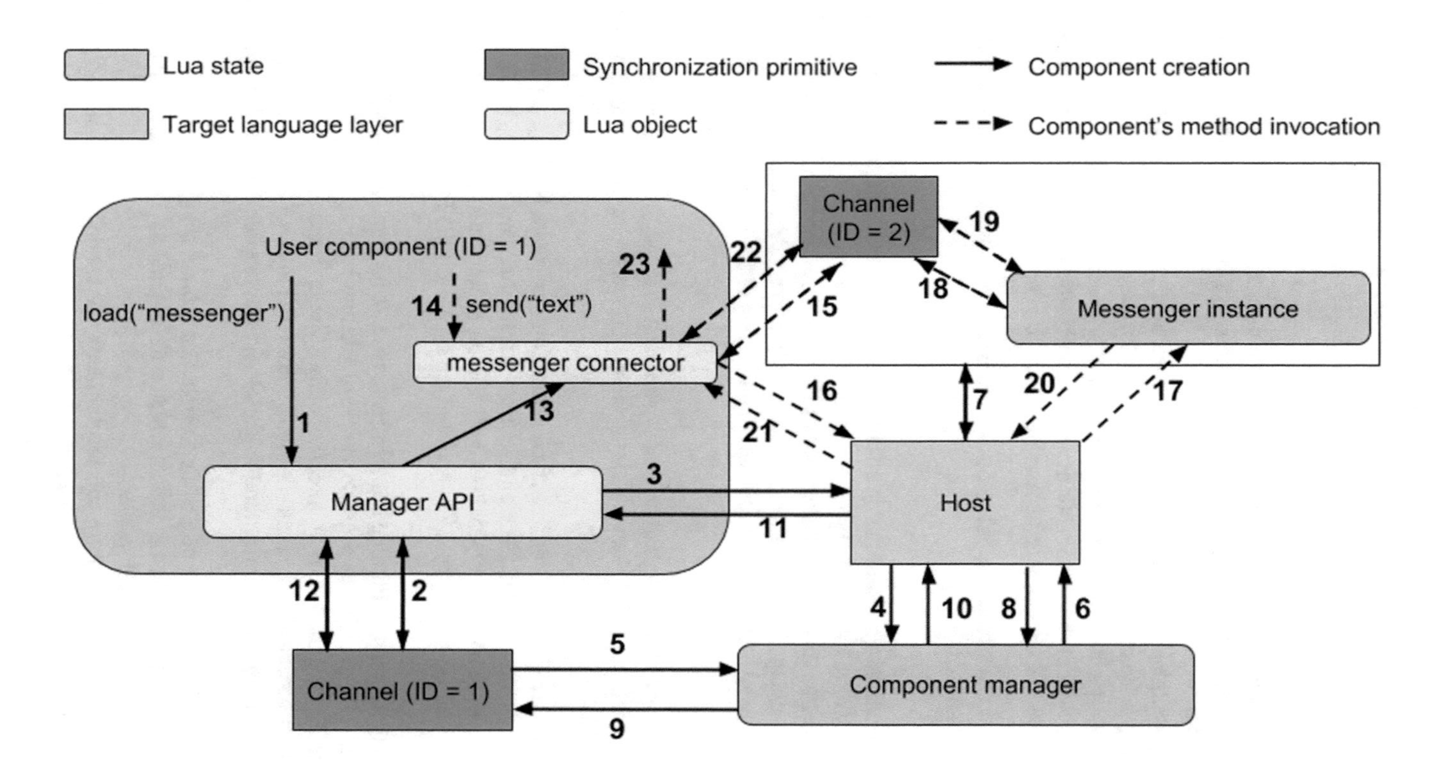

Fig. 2.2 Out-state integration

9. Component manager puts component's connector back to the channel. Connector also includes a unique ID of the channel to communicate with corresponding service component
10. Function returns to step 4
11. Function returns to step 3
12. API Manager reads the returned data from the channel
13. API Manager loads the raw component connector code and returns it (as object) to the user
14. To communicate with a messenger loaded as a service (in a different thread) user pushes data to the channel and waits for the response
15. Messenger checks for message in channel and waits until it appears
16. Messenger performs the actions required and pushes the data back to the channel
17. User wakes up and gets the response from the channel.

Based on the above examples let us summarize the agility pros and cons of the possible solutions:

Pros:

- Component integration logic is encapsulated, the user is able to work with components like regular Lua objects
- Component manager handles the lion's share of the integration activities, so host (target language binding) should provide only a few simple methods
- Channel usage removes the data serialization responsibility from the host.

Cons:

- In case of out-state integration, extra method calls through several layers require synchronization or data locks. This negatively affects the performance of out-state components.

We propose the Common Interface Definition Language (CIDL), a DSL based on Lua. It defines the interface and implements the intermediate logic. The CIDL includes communication types, interface definition functions, type definition functions and Lua syntax support. For an interface definition the DSL uses its own type system.

The target language binding is implemented as an additional module of the system. Therefore, it is possible to override, delete or add any binding implementation. The language binding module is simple; it exports a single function out.

Such architecture involves an additional interpreter layer of communication logic. This layer should unify the communication logic and the interface, and simplify the solution.

The DSL decreases the cost of the system and promotes agility.

The internal DSL is based on Lua which was selected as a simple language well suited for embedded development.

To support the operating system-level threads, we created a multi-threading support for Lua with a high level interface.

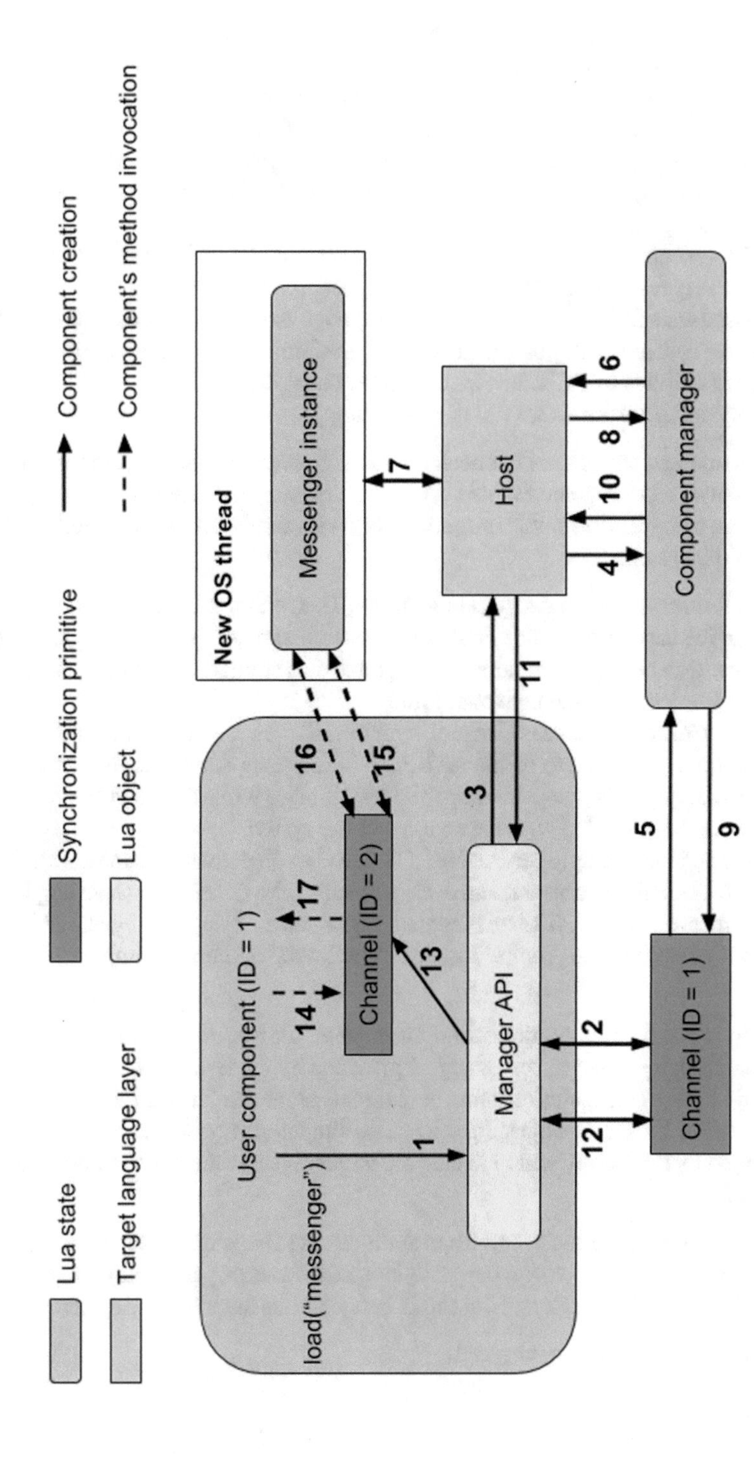

Fig. 2.3 Service integration

Our component system simplifies the use of threads and multiple Lua states. It defines several integration schemes for the components, for in-state, out-state and service. Depending on the integration scheme the component may be loaded into the same state, into a different state and executed in a different thread. The component system loads all required components into a single precompiled Lua component storage. This storage increases reliability, accelerates loading, and keeps the system consistent and agile.

The CIDL supports a set of standard types, allows to define interfaces, functions, complex data types and declare user defined types.

The proposed solution suggests a target language binding as an external module. We keep this layer as thin as possible to simplify the binding and to move the execution logic responsibilities to CIDL level, i.e. to promote agility.

To analyze the outcomes we did the following:

- Smoke, unit and stress tests to check out the entire system and its components
- Performance tests to compare the system with existing solutions
- Extension examples to demonstrate advantages of modular architecture and custom parts of the system.

Smoke tests cover the target system features [7]. The stress test checks the system behavior in the case of multiple clients with multiple threads and in situations when server side interface implementations use a global data storage. The stress test checks the system for deadlock and race conditions.

For performance we created several simple tests. As a benchmark we selected GRPC technology [8], a state-of-the-art RPC solution. This solution is an open source and features multiple language binding. The tests that we wrote for GRPC were based on C++ examples provided with the sources of the system.

Our test bed had a single server with a single C++ object which handled all connections. We called this object a *slave singleton*. For the GRPC, the slave singleton was implemented in C++. The test created a new slave object and registered ID in the GRPC system. For the proposed system, the slave singleton was implemented as follows:

1. **Single-state**: a configuration using an in-state integration scheme and the singleton property of target server object provided by the testing side
2. **Multi-state**: a configuration using an out-state singleton integration scheme and responsible for target object loading. The singleton property in this case was provided by the factory which loaded the target object in the out-state integration scheme.

Multi-state configuration promotes usability as the singleton property implementation is encapsulated. However, out-state integration means performance overhead, because it requires four data transmissions between Lua state and the system:

1. The caller pushes data to channel
2. The callee reads data from channel
3. The callee pushes response to channel

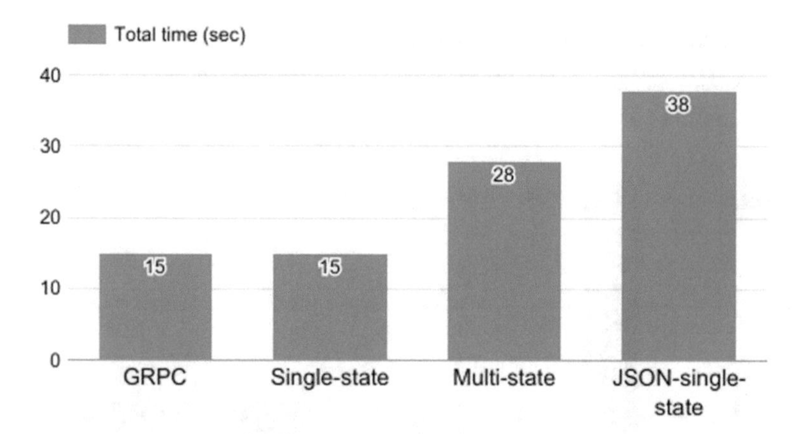

Fig. 2.4 Single-state performance results

4. The caller reads data from channel.

One more configuration, JSON-single-state, was exactly the same as single-state configuration except that it used a JSON protocol for data serialization.

For the following tests, the communication interface and the server side were implemented in the same way. The interface had a `send` method; this method received a string and returned a string. The client created an instance of interface object and called this method repeatedly. For each method call, the client gave a randomly generated string and expected the server to respond with the same string concatenated with a prefix.

A single-connection test checked the efficiency of the network communication. A client application created a single instance of the communication interface and repeatedly produced 100 K requests to the server application. The test measured the total communication time and memory consumption by each process.

In single-state communication (Fig. 2.4), the performance of the proposed solution did not degrade, while the other configurations did. Multi-state configuration showed that the out-state integration slowed performance by almost twice. The JSON-single-state had the worst performance due to slow Lua regular expression implementation. Therefore, we decided to refrain from using it for the rest of the tests.

The memory report (Fig. 2.5) revealed that the proposed solution required less memory for a similar scenario. This meant that the proposed solution was acceptable.

The multiple-connection test was generally similar to the previous ones. It simulated 10 concurrent clients each of which created 5 communication interface instances and used them in different threads (5 threads per client) to send 10 K messages to the server. The total amount of connections was 50 and the total number of requests to server was 500 K.

The performance of the multiple-connection test was comparable to that of the single-connection test (Fig. 2.6). The GRPC and the single-state configuration of

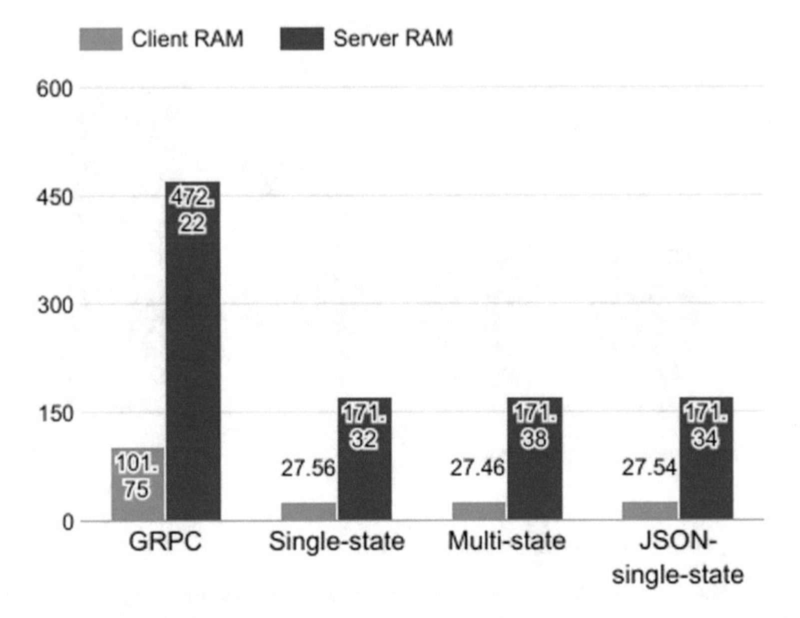

Fig. 2.5 Single-state memory consumption

the proposed solution worked a little faster. The single-state configuration handled about 20 K requests per second which was acceptable; however, the performance of multi-state configuration degraded. The multi-state configuration was almost three times slower than the GRPC, which was an expected result. Due to the structure of the singleton implemented by the out-state integration and multi-state configuration, it was less scalable than the singleton implemented by the target language and single-state configuration (see Fig. 2.7).

The out-state integration required extra data serialization/deserialization. This is why the out-state integration inhibited scalability and agility.

The complex data test (Fig. 2.8) checked the efficiency of data serialization and deserialization. Previous tests used simple plain text but this test used more complex messages. Each message consisted of 5 structures (declared using structural type of CIDL and used message type of protobuf in GRPC). Each of these structures had several fields and may have had a nested structure, the deepest level of nesting was 3. The structures of the last level consisted of two primitive fields. So, the total amount of primitive fields was 20. In general, this test was based on a single-connection test and established only one connection used in one thread to transmit 100 K requests.

Complexity of messages expectedly increased the communication time. Each request's data had to be marshaled to and from the C++ object, so, when the data became more complex the time of serialization significantly affected the performance of the system.

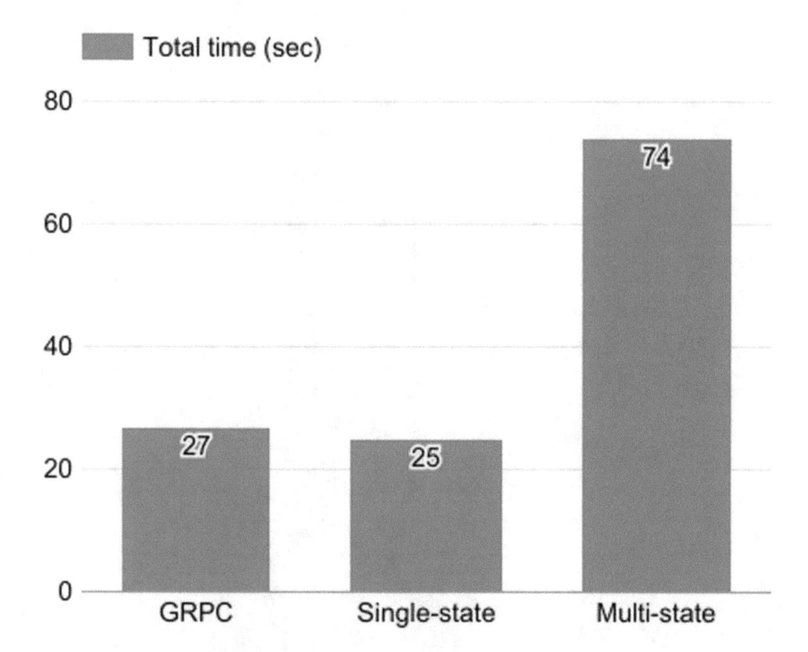

Fig. 2.6 Multiple-state performance results

Performance tests showed that system performance was acceptable. As expected the proposed solution used an interpreted layer comparable to its opponent. The key purpose of the performance tests was to demonstrate overall system utility.

The tests showed that the weak point of the system was an overhead for communication data parsing. This problem was not an architectural flaw but rather a drawback of the current implementation, and this is manageable.

Smoke tests showed that the system worked as expected. Stress tests checked the system under pressure and empirically proved its reliability.

Performance testing revealed several ways to improve the system's performance: switching to LuaJIT, using an `effil.table` directly instead of regular Lua tables or a stack-to-stack copy without data serialization.

In general, the performance tests showed that the interpreter layer of Lua language did not significantly decrease system performance. This proved our concept though generally the performance decreased slightly as agility improved significantly.

The system extension showed the applicability of the modularity principle. The system became easier to customize and extend. Based on the Web server, the users became able to override the communication logics, use the system in different environments and applications. The new custom type, array, showed that system agility improved in many aspects including the interface.

This case study was devoted to RPC-based software development for inter-process communication. The concept of the proposed solution was based on the principle of

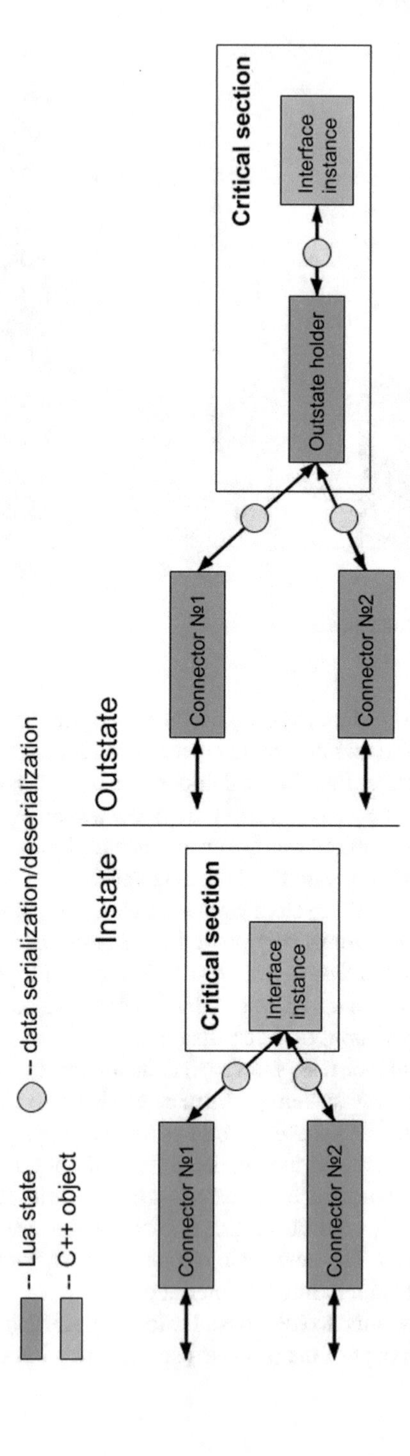

Fig. 2.7 In-state and out-state integration schemes

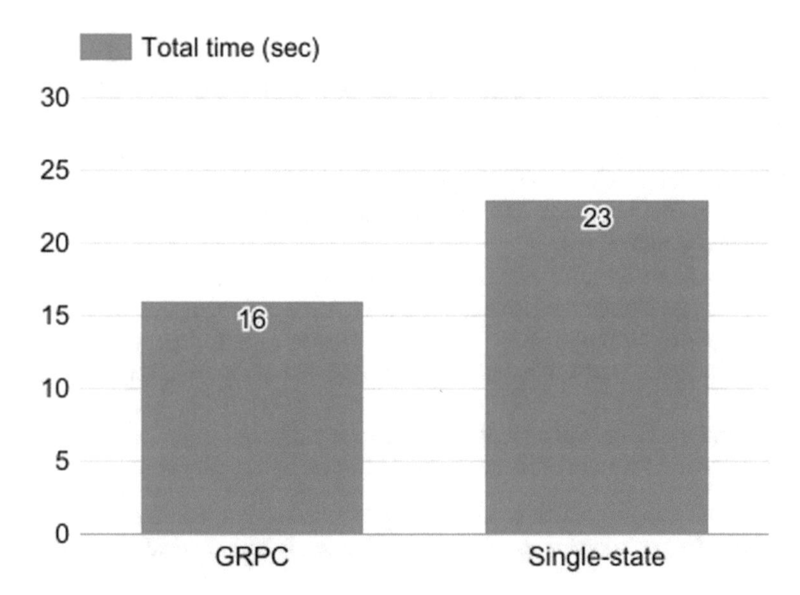

Fig. 2.8 Complex data performance results

an intermediate interpreted layer. This layer simplifies the application of technology, unifies the development process and improves code (re)usability.

The interfaces are defined using a DSL which provides a list of standard types for data transmission and a set of keywords for interface structure and data description. The DSL is extendable and allows the user to create complex data types using known types, and define custom data types. The DSL is of internal type and based on the interpreted language of the intermediate layer. This principle simplifies the interface definition by means of a DSL and maintains full functionality of the base language. As a base language, the solution uses Lua because of its efficient embedding facilities, simple syntax and high performance.

Our open-source project, Effil, provides a multi-threading support for Lua and a high level interface for inter-thread communication.

Each component provides a *manifest* (i.e. a metadata file) which declares how the component should be loaded and which public interface the component implements.

To manage the component storage, we developed a command line tool. This tool generates the source code for the application language. This source code is a connector between the layer of endpoint applications written in target language and the interpreted layer of the proposed solution. This generation process is a responsibility of the target language binding. The language binding is implemented as an external module for user-defined binding implementations. The system provides single binding for C++ language. The C++ binding of the developed system supports customization.

The inter-process communication logic is implemented by a set of components organized as a component storage. These components are responsible for data serialization, network connections and server instance support.

To prove the concept, we developed samples and tests that demonstrated better utility and agility. One sample implemented a server side HTTP connector which allowed using the proposed RPC framework as a Web server. The other sample implemented a custom type.

Testing included smoke tests, stress tests and performance tests. Test results showed that the solution performance was acceptable. Generally, the solution based on the interpreted layer performed adequately. However, performance tests identified a problem in the current implementation: an overhead of data exchange between the Lua states.

In general the result was an agile inter-process communication framework with a customizable modular architecture. Ways of further agility improvement are:

- Add LuaJIT support which can increase performance 2 to 60 times
- Use stack-to-stack copy to avoid data serialization in communication between Lua states
- Implement bindings for other languages. Lua already has bindings for: Java (LuaJ), Python (lupa), Go (go-lua), Rust (rust-lua), C# (MoonSharp) etc.
- Create components to extend the functions of the system, e.g. request caching, and binary protocol
- Create new data types for communication.

2.4 Developing an Embedded System

There is a growing demand for rich user interfaces in embedded systems which have been influenced by the arrival of touchscreen smartphones, complex software systems, and embedded systems. Implementing complex user interfaces in the embedded system requires a great amount of labor. Therefore, low level tools appeared to automate embedded system development. However, most of these tools are too generic, and the code generated from their models is manually customized in general purpose languages. To develop rich interfaces for embedded systems, it would be more effective to have a language that is portable, maintainable and flexible. This would reduce complexity and improve agility of these, often mission-critical, systems. We suggest developing a projectional DSL-based user interface for embedded systems.

Embedded system development has improved through the years with the advent of touchscreens and complex user interfaces in mobile phones. These kinds of systems are complex, and often mission-critical for business, industry, and daily life. Embedded systems exist in most electronic devices from consumer electronics (smartphones, TV, microwave etc.), to industrial and military systems.

Embedded system development depends on strategically reusable components, which provide agility due to variation in product lines. Traditional development of a user interface for embedded software systems requires: (i) designing a visual interface, (ii) writing the code for the interface, and (iii) testing the code in a certain environment.

Embedded software interfaces have grown from simple buttons, LEDs and control panels to sophisticated user interfaces.

DSLs assist in solving problems within a specific domain. The *domain specific modeling languages* (DSMLs) are typically concise and convenient, they improve model quality and comprehensibility [4]. In DSM, the solution starts from the idea of the problem domain. This sketch contains the syntax and semantics that represent the concepts at the same level of abstraction that the problem domain offers.

Language-oriented programming (LOP) is a software development approach that utilizes DSLs and DSM to solve a problem in a specific domain [9]. LOP saves time for writing code, since the coding starts from the mental model in the DSL, and therefore potentially promotes agility.

Recent DSLs feature customizable modeling languages and code generators. These tools support modeling environments. They provide user-friendly GUI front-ends for meta-modeling and/or code generation (often as a toolbox with visual diagramming). The DSL vendors developed *language workbenches* (LW) to help software developers automate development processes. These tools have graphical and/or text notations to represent the DSL concepts. The models of these LW compose UML structure and behavior diagrams enriched by manual code in a "regular" programming language. This complicates product debugging, documenting and maintenance as its components are loosely coupled.

Another drawback of many DSL tools is the loss of extensibility which occurs after transforming the models into code. Language workbench agility is also greatly hindered by their inability to incorporate non-parsable notations (such as tables and images etc.), which, however, are often useful in embedded system GUI.

We propose to use a *projectional* kind of editor to develop a GUI generator language for embedded systems. This prototype will help to develop and evaluate embedded system models from a high level mental model. We will create a proof-of-concept by developing a DSL prototype, and a plugin for a standalone IDE.

In the lifecycle of complex software systems, particularly enterprise ones, many routine deliverables are generated. Traditional software development often starts from scratch and requires manually writing code, which sometimes seems inefficient. Even recent software engineering tools have a complex support for data/language extensions which limits their applications to a specific domain [10].

The Microsoft DSL Tools are available for use in the Visual Studio (VS), the user needs to install the plugin before its features become available. This tool enables language designers to build DSL framework and then generate modules or artefacts required for the language. With these tools, language designers can create, manage and visualize models that are underpinned by a code framework, which allows defining domain specific schemas, and then constructing a custom language hosted in MS VS [11]. Although much effort is needed to learn how to use a DSM tool and develop a DSM solution, MS DSL Tools requires less effort in learning and creating a DSM solution, it involves mostly dragging and linking shapes and diagrams together.

Most times, this could also involve adding some code if the developer chooses to add semantic validation separately. During code generation, MS DSL Tools generates proprietary C# code which can be further enhanced only with C#. This process of enhancing the language features leads to a loss of agility of the DSL created and makes the language hard to maintain. Language extensions and combinations are impossible, which, again means lower agility.

The Sirius Obeo Designer provides support and collaborative features for DSL development. Languages and tools developed with Obeo Designer usually comprise of a set of editors (e.g. diagrams, trees etc.) which allows defining and modifying models in a shared repository. Obeo Designer utilizes Eclipse technologies (Ecore, EMF, and GMF) to create graphical DSLs. In terms of graphical completeness, Obeo Designer offers better agility than MS DSL Tools as it is based on viewpoint representation i.e. a set of descriptions to view domain concepts depending on the user's role. Obeo Designer relies on EMF Ecore and thereby requires that the domain experts create models in Ecore before importing them into the project. Structural codes generated from their UML models are manually enhanced, which also means a loss of agility.

Graphical Modeling Framework (GMF) is an open source Eclipse Modeling Framework (EMF) used for creating DSL. It comes with a cheat sheet which serves as a guide for DSL developers [12]. GMF consists of three parts: notation, runtime and a tooling framework. The notation provides a EMF standard meta-model that separates and retains the diagram information from the domain model. The runtime handles the task of bridging EMF and GMF while providing a number of services and an API for external graphical editors. The tooling component provides a model-driven approach to describe graphical elements, to diagram, and to generate diagrams for the runtime [13]. It uses the EMF Ecore model file as input, from which it derives the templates for the graphical editor. GMF creates a generated graphical editor for instances of the meta-model [14]. GMF relies on EMF Ecore, which inhibits agility, as its models must be Ecore-based. Another limitation is that the codes generated from the UML models must be manually enhanced.

Let us compare the above DSL tools based on the key criteria identified below.

Graphical Completeness: Obeo Designer offers more agility in this category as it is based on viewpoint representation [15]. MS DSL Tools is the second in terms of agility for domain model representation and supporting GUI. GMF is even less agile.

Efficiency: MS DSL Tools requires less effort in learning and creating DSM solution due to dragging and linking shapes and diagrams. It allows adding code for semantic validation. Obeo Designer and GMF rely on EMF Ecore which is a clear agility limitation.

Support/Tool assistance: GMF comes with a cheat sheet to guide DSL development. All the three tools have basic help systems and online user community resources.

Code Generation: All the three tools support code generation. Of these, MS DSL Tools produces proprietary C# code, which probably means less agility.

Diffuseness: It takes fewer steps to produce a model in Ecore; however, transforming this model into meta-model in both GMF and Obeo Designer requires more effort. MS DSL Tools is more agile as it does not require an external model to create its meta-model.

Reusability: Obeo Designer and GMF rely on EMF Ecore, so reusing a model or meta-model is their basic feature. MS DSL Tools also supports a sufficient reusability level.

Licensing: Microsoft DSL Tools requires commercially licensed Microsoft Visual Studio platform while GMF and Obeo Designer are more agile as they are available as freeware under Eclipse Public License [18], and have options for commercial support.

Maturity: All the three tools are used for industrial production of complex software, i.e. applicable for enterprise applications.

Maintainability: These tools have a relatively low maintainability since it is difficult to extend them by enhancing their core languages.

Generality: All of the tools evaluated are generic, i.e. applicable for a wide range of domains. The goal of DSL is to target a specific problem domain; therefore their generality may lead to a deficit of agility.

Extendibility: A DSL should be portable and easily integrated/blended with an existing language. The tools examined cannot be easily extended by a rich set of other languages. As an example, Microsoft DSL Tools code can only be enriched by C# code.

Enterprise experiences with DSL/DSM show significant improvements in development productivity, costs and product quality as the solutions are based on problem domain models rather than on code models [15, 16]. For instance, DSL-based development of Panasonic Lifinity touch screen controller for lights, heating, air conditioning etc. provides 300–500% productivity increase as compared to the manual approach [15]. Another application is Sport Computer by Polar, a product that measures, analyzes and visualizes data on heartbeat, calories, speed, distance, altitude changes, pedaling rate etc. [15]. The product features depend on the kind of sports, such as cycling, fitness and team games. The DSL by Polar covers the GUI only. The influence of DSL/DSM was evaluated by Polar: (i) by building a large portion of the product in a pilot project, and (ii) by asking six developers to individually implement a small typical feature. The development time with ordinary versus DSL approach was: (i) 23 days versus 2.3 days and (ii) 16 h versus 75–125 min, i.e. nearly 900% productivity improvement with DSL.

Embedded systems require a careful design so that their GUI is usable, intuitive and efficient. In addition to usability [17], we consider a few other factors: robustness, consistency, compatibility, control and responsiveness. Importantly, these are agility related factors.

A GUI interface is *robust* in how it can respond and handle unusual user input while permitting users to customize the interface.

A *consistent* GUI ensures that a user can easily understand how it works in general thus ensuring smooth transition between different areas of the GUI. As noted in [17], inconsistency in the GUI results in errors injected into the code.

Compatibility is matching the function the user interface performs with users' expectation of the system, that is, GUI should match user's prior knowledge of the system software and hardware (we describe prior knowledge in more detail in [2]).

Responsiveness means that the feedback to the user must be informative and well acknowledged (we describe feedback in more detail in Chap. 4 and in [2]).

In the case study, we use a projectional editor from JetBrains MPS. Projectional editors directly manipulate the abstract syntax tree (AST).

We derived the name *ESUILang* from the abbreviation of "Embedded System User Interface Language" which represents the implementation of a user interface for embedded system from language features and code generated from the users' high level mental models. ESUILang is made up of different components and each of these components integrates together to ensure the language works as a whole.

To develop the embedded system GUI generator language, ESUILang, let us set the key requirements.

Based on our analysis of existing DSL tools, we believe that ESUILang should combine certain features of the tools reviewed and add some more features in order to improve the agility of the target embedded systems.

We believe that the ESUILang should have the following features: (i) domain focus for embedded system UI, (ii) grouped constructs; (iii) smart coding; (iv) instant error checking; (v) visible code representation and execution; (vi) industry standard compliance.

Single Domain Focus: The language is targeted for a single domain of the embedded systems; this makes the language compact and easily extendable.

Extensibility and Composability: The language should be portable and available as a plugin which simplifies embedding and allows inheritance.

Support for Custom Notations: The language should support non-parsable notations (e.g. tables, mathematical symbols, equations and expressions) which become increasingly significant for embedded system development.

Open Source: ESUILang is available as a plugin to extend existing IDE and the code is also available on git.

ESUILang (Fig. 2.9) is built on top of the Jetbrains MPS projectional editor.

ESUILang permits developers to combine built-in GUI elements with custom implementation in another language. The implication of this is that a developer can either use only the minimally required features of ESUILang to develop their application or implement the user interface development entirely in ESUILang.

The ESUILang consists of several language constructs. Each of the constructs is made up of one or more concept which in turn can have another language aspect (e.g. constraint, behavior etc.) A language diagram is generated from the language core. This language diagram shows all the concepts in the language core and their dependencies.

UI language concepts in ESUILang consist of shapes, diagrams, and relationships. ESUILang improves the structure of the application, and supports different implementations of the same interface. This improves modularity and testability, which are agility attributes. The component language core is used for expressing dependencies

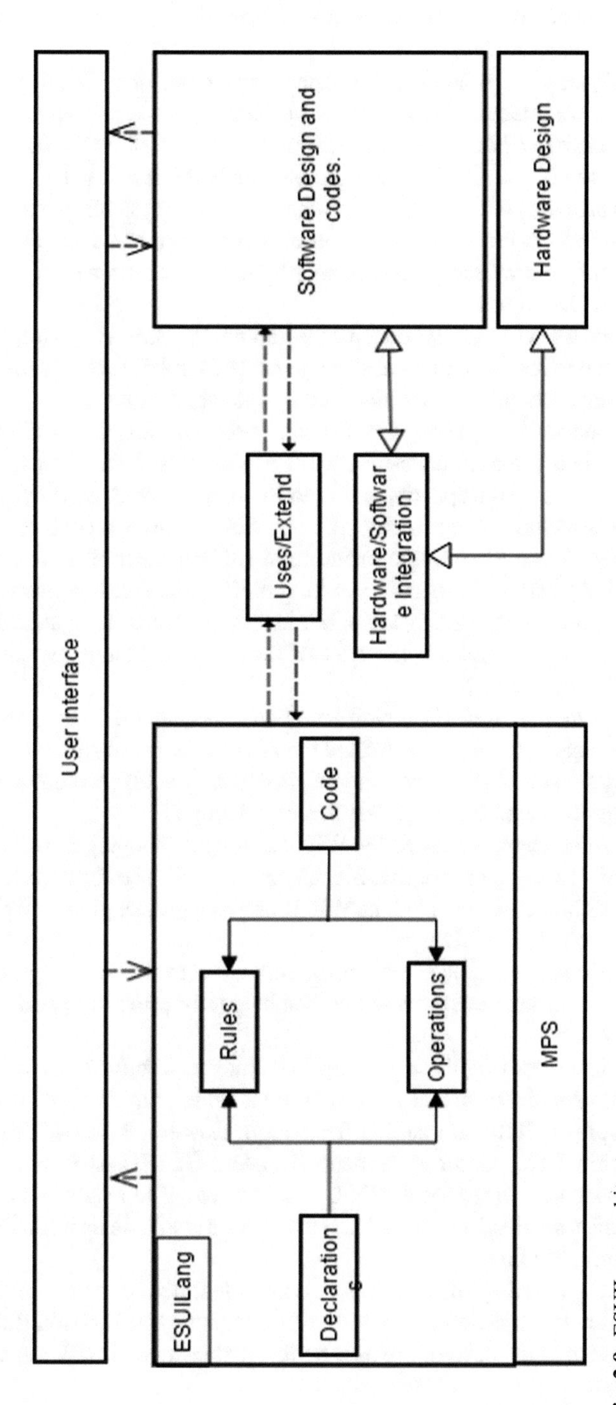

Fig. 2.9 ESUILang architecture

between components of a system, which provides textual, tabular and diagram editor notations.

Embedded systems communicate with the environment via different protocols. To improve agility, we refactored one of these protocols to embed a state machine. We combined the state machine and the other components to allow for the decoupling of message assembly and application logic parsing at the server side.

To develop an agile GUI for mission-critical embedded systems, we need to create a language that reduces implementation complexity and improves quality. ESUILang design is modular, and we reduce complexity by means of certain composition techniques aimed at DSL reuse.

The structural aspect contains concepts that define AST structure of any ESUILang code. We created at least one main abstract concept (named `conceptNameType`) in each language core; this will be inherited by the other concepts.

Abstract (concept) type is a common ancestor for all statements and `Expression` is a common ancestor for all expressions. Behavior aspect defines methods for nodes and concepts that are used in other aspects, e.g. every expression must implement method `TypeAnnotation`, which returns an expression's type. A concept in MPS can have single or multiple editors. Each of these editors may have a tag called an editor hint, and by setting hints in the editor window, users can switch between notations. Models are either text or diagrams. Language diagram in ESUILang (Fig. 2.10) represents the AST of the individual language concepts that made up ESUILang.

A model-to-model transformation method was applied for transforming ESUILang concepts into code; we defined the AST node transformation. ESUILang is an open source solution, which is agile. ESUILang is also available as a cross platform plugin to extend language features of existing IDE.

As a proof of concept, we tested the ESUILang by developing a Robot language extension. This is a virtual robot that moves objects from one location to the other. To extend the Robot, we embedded the ESUILang as a project plugin. We improved the user interface with ESUILang.

Adopting a domain specific approach for embedded system development can not only speed up the production process but also help to improve software quality and the user experience.

We started by comparing three existing DSL tools and found that these tools are either incomplete or inconsistent in certain areas which are critical for embedded system development (Table 2.1 summarizes our outcomes). Based on this, we identified the key requirements for an embedded system GUI. These key requirements became our basic guidelines for the DSL development. The result was a prototype of a DSL for an embedded system GUI. This DSL is available as a standalone IDE and in ESUILang Studio.

ESUILang improved agility in terms of adding features for user interface development and allowing to develop an intuitive and scalable UI. The ESUILang Studio was customized to allow developing user defined extensions; ESUILang can also be used as a plugin.

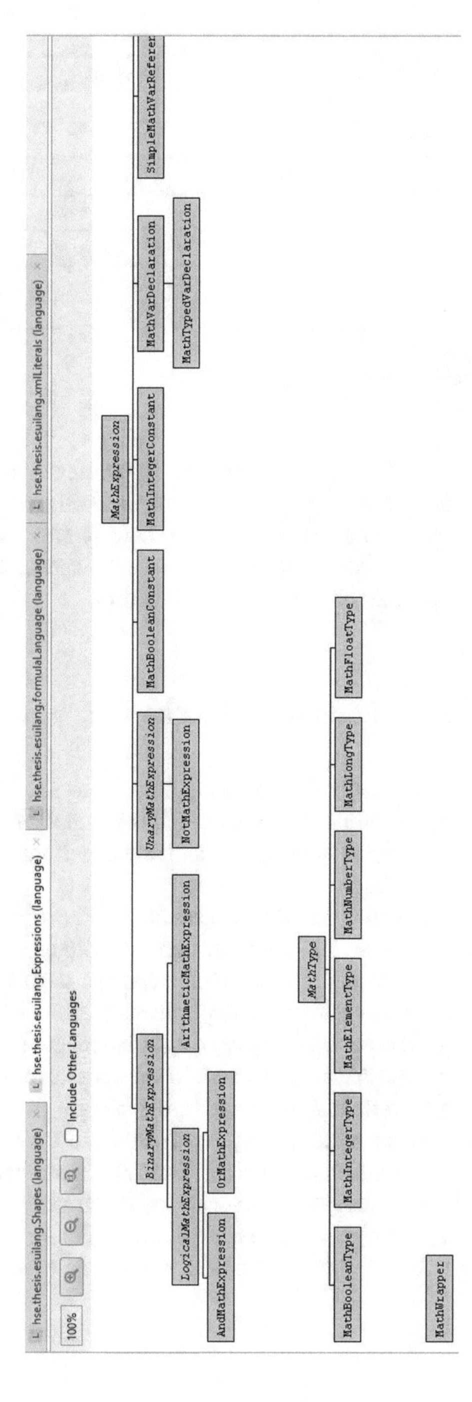

Fig. 2.10 ESUILang expressions diagram

Table 2.1 ESUILang in comparison with other tools

	MS DSL tools	Obeo designer	GMF	ESUILang studio
Effort	Low	Medium	Medium	Medium
Code Generation	Yes	Yes	Yes	Yes
License	Commercial	EPL	EPL	Apache 2.0
Maturity	++++	++++	++++	+
Maintainability	-	-	-	+++++
Non-genericity	-	-	-	+++++
Language Composability	-	-	-	+++++

"+++++": represent a positive feature. More is better
"-": denote areas where the feature is missing

To further improve agility, we recommend: (i) adding a gesture language; (ii) adding a hybrid editor for Web application of the DSL; (iii) including code analysis for the language workbench to improve security and reduce complexity. These recommendations will specifically improve user experience and developer productivity for ESUILang and other projectional editors.

2.5 Conclusion: How Languages Work

The previous chapter created the framework for agility, crisis and lifecycle including their interrelations, resilient adjustment and responsive application. This chapter added the interface aspects for the applications and processes that enhanced the agile approach by means of a language.

We classified the programming languages, presented their evolution and examined several types of general purpose languages in light of agile development of large-scale software applications. Clearly, no universal remedy exists in terms of languages for crisis development of large-scale software products. However, our first finding is that out of the old-fashioned dinosaurs only most agile languages survived. To further improve agility, the crisis resistant software should incorporate the components written in the language paradigms that are most appropriate for the problem domain. This approach improves scalability, modifiability, portability and overall maintainability (i.e. a number of agility dimensions) of the mission-critical software products. The ultimate application of this strategy to large-scale agile software development is the DSL, or domain specific language, approach. To illustrate this strategy, we presented case studies on inter-process communication and embedded system development.

However, this strategy does not provide ultimate agility yet. To improve agility even further, we must better align the lifecycles and languages with the business constraints and technical requirements. These extra dimensions for agility fine tuning include human factors, design patterns and best practices. We are going to discuss them in the later chapters.

References

1. Pratt, T. W., & Zelkowitz, M. V. (2001). *Programming languages: Design and implementation* (4th ed.). Upper Saddle River, NJ, USA: Prentice-Hall, Inc.
2. Zykov, S. V. (2016). *Crisis management for software development and knowledge transfer issue 61: Springer series in smart innovation, systems and technologies* (p. 133). Switzerland: Springer International Publishing.
3. Blake, D. (2007). Domain specific languages versus generic modeling languages. Retrieved November 07, 2017, from http://www.drdobbs.com/architecture-and-design/domainspecific-languages-versus-generic/199500627.
4. Chappell, D. (1998) The trouble with CORBA. Retrieved November 8, 2017, from http://www.davidchappell.com/writing/article_Trouble_CORBA.php.
5. Van Deursen, A., Klint, P., & Visser, J. (2007). *Domain specific languages: An annotated bibliography*. Amsterdam: The Netherlands.
6. Ierusalimschy, R. (2003) Programming in Lua.
7. Dustin, E., Rashka J., & Paul J. (1999) *Automated software testing: Introduction, management, and performance*. Addison-Wesley Professional.
8. GRPC: A high performance, open-source universal RPC framework. Retrieved November 7, 2017, from http://www.grpc.io/docs.
9. Dmitriev, S. (2004) Language oriented programming: The next programming paradigm. Jet-Brains, 1–13.
10. Voelter, M. (2014). Generic tools, Specific Languages. CreateSpace.
11. Furtado, A. (2002) Tutorial: Applying domain specific modeling to game. Microsoft Innovation Center at Recife/Informatics Center (CIn).
12. Gronback, R. (2009) *Eclipse modeling project: A domain specific language (DSL) toolkit*. Addison-Wesley.
13. Gouyette, M. (2010). GMF graphical editor tutorial: How to create a FSM graphical editor with GMF?
14. Sirius. (2017). Modeling projects and representations. Retrieved November 7, 2017, from http://www.eclipse.org/sirius/doc/user/general/ModelingProject.html.
15. Tolvanen, J.P. (2014). Industrial experiences on using DSLs in embedded software development. Retrieved November 7, 2017, from http://metacase.com.
16. Kelly, J.P.S. (2008). Enabling full code generation. In *Domain Specific Modeling* (p. 44). Wiley, IEEE Computer Society Press.
17. Murphy, N. (2000) Principles of user interface design. In *Embedded Systems Programming* (p. 55).
18. Eclipse. (2017). Graphical modeling project. Retrieved November 7, 2017, from http://www.eclipse.org/modeling/gmp/.

Chapter 3
Agile Services

Abstract This chapter describes the best practices for enterprise software development. We discuss Microservices and how they improve organizations by providing more security and further efficiency benefits; we examine how service-oriented architecture (SOA) and Microservices differ from the monolithic approach and where they are applicable. Another section of this chapter is dedicated to the continuous delivery and continuous integration of the software development lifecycle. We present a detailed case of Microservices in the banking sector and investigate how these help to meet the rapidly changing business requirements. Further, we analyze customer relationship management integration with geo-marketing and their synergistic effect on the business. Finally, we discuss cloud services and particularly virtual machines and test how they help improving enterprise agility.

Keywords Enterprise software development · Microservice
Customer relationship management · Banking · Cloud service

3.1 Introduction: What Is an Agile Service?

The previous chapters outlined the concepts for agility and crisis, lifecycle and languages. They also discussed how the agility level can be improved, and how lifecycle can be tuned by means of special purpose languages in order to become crisis resistant. However, this part of the story was generally about the developer, whereas the client's part was either high level or out of the scope. As we said, this client's part of implementation was possible by means of a language-based interface. Still we gave little details about how this interface might look like.

This chapter describes the principles and practices of different kinds of service-based approaches and their application to software development. We discuss several types of service-based architectures such as service-oriented architecture (SOA), grids, clouds, Microservices, and a few others. For each type of service-based approach, we analyze its strong and weak sides from the standpoint of agility. Thus, the aim of this chapter is to provide recommendations on adjusting certain types

S. V. Zykov, *Managing Software Crisis: A Smart Way to Enterprise Agility*,
Smart Innovation, Systems and Technologies 92,
https://doi.org/10.1007/978-3-319-77917-1_3

of services so that they can be used in crisis conditions and/or for mission-critical software development.

The early service-based approaches were addressed by IBM in their initial ideas of SOA, which later evolved into Software-as-a-Service (SaaS) architecture. The subsequent development of the concept gave rise to grid computing (proposed by Foster et al. [1]), which substituted supercomputers for thousands (and sometimes millions) of ordinary network-connected PCs. As we can see now, all the above architectures aimed at management of extremely large and rapidly increasing volumes of data, or what we now call the big data. To a certain extent, these service-based approaches were a remedy for the crisis of big data management, which could be only partially solved by means of databases and data warehouses (an influential conference name popular at that times was VLDB, or very large data bases). This data management crisis of early 2000s was conquered by IBM, Oracle and Microsoft, the software giants and the pioneers in large-scale database management systems (DBMS).

And this victory was based on service-oriented computing. First it was IBM (and the other giants) with the SOA concept. Second, it was Foster with grids. Third, it was Ellison, the Oracle CEO, who proclaimed the era of cloud computing [2]. Next, it was Microsoft to successfully offer for lease the mass-scale software services deployed in a cloud, such as their famous Office365, Outlook, and Windows Live Services. Finally, it was West, who suggested Microservices, which essentially is a concept that provides a higher level of agility [3].

Clearly, our research will conclude that no single service-based approach is a "silver bullet" in crisis resistant software development. However, our findings will help to build and improve client-oriented services so that they enhance the end user software products' agility.

One more valuable outcome is presenting case studies of service-based approaches that include real world examples in different problem domains. Even though we have concluded that certain problem domains require DSLs, the overarching service-based architectures may stay relatively language independent and still yield to an agile product. The problem domain instances for the case studies include banking and finance, customer relations management, and geoinformation systems. We also present a general purpose business application scenario for optimizing resources of virtual machines in the cloud.

The practice of service-oriented development clearly depends upon a number of human-related factors that include business constraints and technical requirements, and that may dramatically help or hinder the final software product agility. The case studies of this chapter only give quick hints on the human factors; a more detailed discussion follows in Chap. 4 and [4].

This chapter is organized as follows. The introduction (Section 3.1) and Section 3.2 overview the evolution of service-based approaches, from SOA to Microservices, and give their informal classification. Section 3.3 examines the Microservices and their application to developing an enterprise-scale software system. Section 3.4 presents a concise outline of a software solution for banking based on Microservices. Section 3.5 discusses Microservices in more detail in terms of their application to developing an agile customer relationship management system.

Section 3.6 analyzes cloud services in general and suggests an approach for improving agility aspects of virtual machines, such as performance tuning and resource optimization. The conclusion summarizes the results of the chapter.

As before, our focus in discussing the services will be their agility for crisis resistant software development.

3.2 The Microservice Approach

In the recent years, Microservices have attracted much attention. What makes Microservices methodology very attractive is that the approach is in fact the description of the best practices which were successfully applied by major companies. Although it began from user-oriented Web applications, now enterprises also understand the efficiency and the benefits of the cloud while modern technologies are able to provide the required security level.

For many reasons, monolithic applications no longer meet the requirements of complex enterprise systems. Despite the fact that well-known SOA provided many brilliant ideas and conceptions, it has not become very popular and widely used because of the complexity and high abstraction level; however, in some way SOA laid the foundation for Microservice architecture.

The application of the Microservices is a very challenging process for many reasons. Dividing the business scope, and further subdividing it by functionality and responsibility makes the software development process inconvenient and excessively complex for the team. Besides, apart from Microservice development, the team should be familiar with continuous integration and continuous delivery. Considering that the system may contain over 10 services, and subsequently, 10 teams and 10 databases, the project and system management becomes complicated and messy. By conducting an analysis of Microservice characteristics, we formed a list of the functions that could be automated in order to simplify the process. One of the main suggested features is the implementation of various patterns. The patterns were analyzed and provided as a part of pre-coded solution. The final part of the study consists of the architectural solution, which includes system architecture, lifecycle, and use cases. Based on this architectural solution, we plan creating a Microservice management platform.

In the beginning, typical enterprise system architectures were monolithic (Fig. 3.1). According to the definition, *monolithic* application implies that functionally distinguishable aspects are not architecturally separate components

but are all interwoven. More specifically, a three-tier architecture pattern may also be described as monolithic as far as all business logic is still implemented using one server, and all data is stored in one database.

Modern applications grow with enormous speed and need to be scalable. That is one of the monolithic application disadvantages—the size of the application causes significant complexity which makes it difficult in development, testing and deployment. In order to make any change the developer should be aware of the whole system

Fig. 3.1 Monolitic application

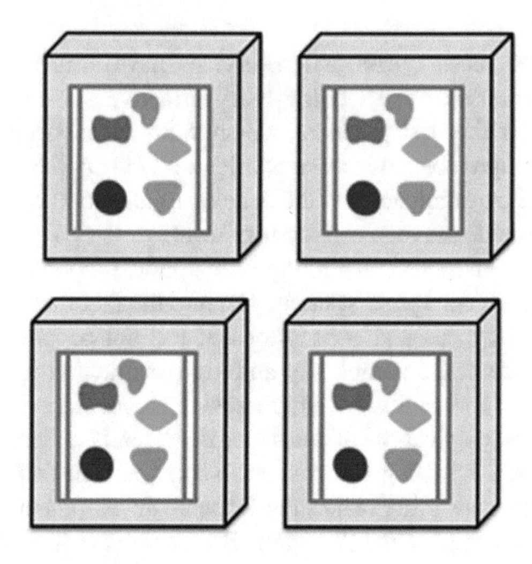

functionality and specific features, the whole system should be replicated during the update. That is very expensive for the customer in terms of human resources and time. Therefore, a new approach was needed.

Thus, three-tier architecture evolved into multi-tier application. One of the examples is the famous Model-View-Controller pattern, which separates the business logic layer, user interface and user interaction functionality (Fig. 3.2). However, business logic in enterprise applications could be extremely significant, so the straightforward separation does not solve the problems. The enterprise applications still remain monolithic.

The methodology that has laid the foundation of the modern network distributed systems is SOA. One of the universal definitions of the term is the following:

SOA is a loosely-coupled architecture designed to meet the business needs of the organization [5]. To be more specific, SOA does not necessary imply the use of Web services, but Web services are the most common modern approach. According to SOA, loosely coupled services are combined to provide the functionality of the enterprise application. Each service represents some specific function. Thereby, the scalability of the system is supported.

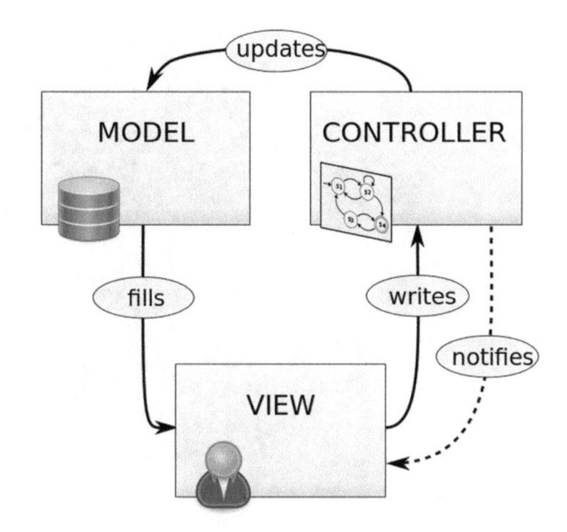

Fig. 3.2 Model-view-controler

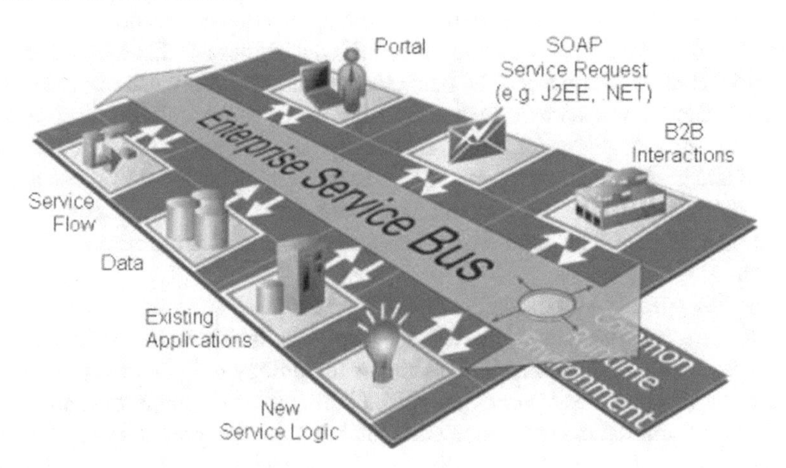

Fig. 3.3 Enterprise service bus

Unfortunately, SOA did not successfully accommodate due to a considerable number of complex abstractions and legacy protocols. The need to connect many services, written in different languages, has led to the complex Enterprise Service Bus which in turn gave birth to the archaic and expensive enterprise applications, which required a lot of resources every time the business scope changed or a new function added [6] (Fig. 3.3).

SOA has brought a very important idea and partly has become the standard for Web applications, but the rapidly changing requirements of enterprise systems demanded a new approach.

"To deal with the rapidly changing demands of digital business and scale systems up or down rapidly, computing has to move away from static to dynamic models.

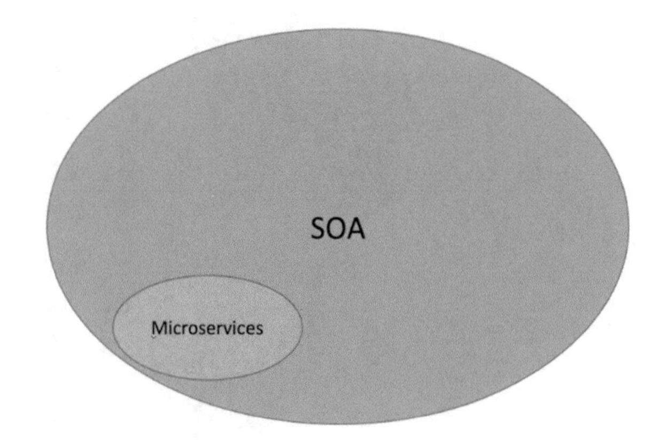

Fig. 3.4 The relationship between SOA and microservices

Rules, models and code that can dynamically assemble and configure all of the elements needed from the network through the application are needed" [7].

Fowler, the pioneer of *Microservices*, initiated numerous discussions, articles and books about this concept. The basic idea of the Microservices correlates with SOA, but Microservices provide a clearer vision. Some people call Microservices the successful implementation of the SOA. Another opinion states that Microservices describe and implement only a part of SOA. SOA is a high level Enterprise Architecture methodology, while Microservices are more about the system architecture and project management. The relationship between SOA and Microservices is shown in Fig. 3.4.

Generally speaking, Microservice architecture implies the separation of the system into different segments of the business scope, so that each Microservice could work as an independent unit. Microservice methodology reduces heavy Enterprise Service Bus and provides clearer endpoints. In order to avoid ponderous integration, which would ruin the independence and scalability, the integration is supported by the usage of HTTP via RESTful APIs, JSON data, and message queues. Microservices are tightly connected with cloud computing, as far as it is absolutely necessary to continuously test and monitor the behavior of the separate Microservices and the system overall (Fig. 3.5).

Fowler pointed out that he did not invent a new architectural pattern or style, but generally summarized and formalized the best new practices and described the most distinctive features of this trend [8]. The pioneers of the style's successful implementation include Amazon, Netflix, and The Guardian.

While discussions and articles have grown into books, the leaders of modern Web application technologies suggested the new features for more convenient Microservices implementation. For example, Spring has announced Microservices with Spring Boot and Spring Clouds.

One of the key features of Microservices is decentralization. Each Microservice has its own team, internal structure and database. In order to perform and control the continuous operation of the entire system, cloud computing tools (such as Travic CI,

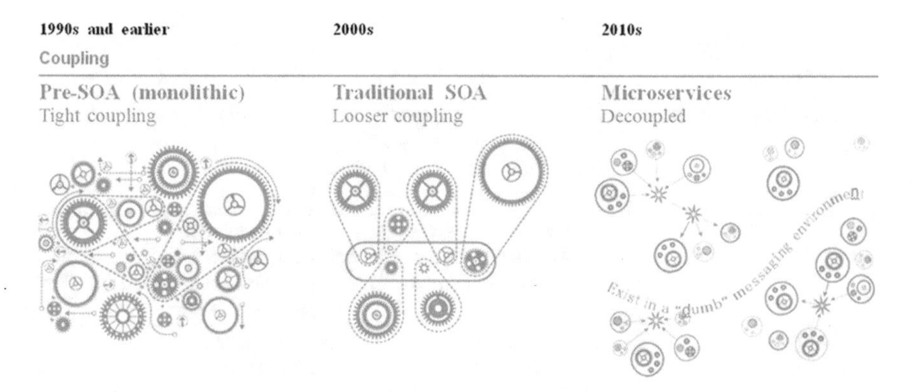

Fig. 3.5 Architecture timeline

Docker, CodeCov etc.) are used. However, such tools should be embedded into the system or coded manually or via API. Most probably, the output information would be accessible to the developers or team lead. Therefore, periodical maintenance is required.

The possibility to automatically track the progress of each Microservice and the system overall, gather statistics and metrics, check graphics and charts, would provide an effective management tool. In addition, embedded integration and design patterns, continuous integration and delivery tools would provide the multi-functional platform for Microservice system implementation and management.

3.3 Enterprise Microservices

The specifics of the enterprise applications are the demanding security requirements. A few years ago it was impossible to imagine that enterprise activity data could be stored outside the internal network. However, with the growth of cloud technologies, enterprise applications have become more cloud-oriented.

There are several options to use clouds for enterprise applications.

Enterprises may use Private Cloud which implies the storage of cloud technologies in a local data center. However, performance similar to that of the Public Cloud would be very expensive. Public Cloud often provides Private Cloud functionality, which considerably reduces the cost. The most desirable approach is to use a Hybrid Cloud which implies the integration of local and cloud applications. Table 3.1 demonstrates that clouds support faster and cheaper enterprise applications than the traditional approach.

SaaS, PaaS and IaaS In the scope of cloud services, enterprise may choose the following facilities (Fig. 3.6):

Table 3.1 Traditional architecture versus clouds

Characteristics	On-premises	Cloud
Capacity and performance	1. Depends on the hardware 2. Maximal performance load should be beforehand planned and prepared	Unrestricted
Cost	1. High hardware cost 2. The same cost for different performance metrics	1. Lower cost 2. Cost depends on the traffic 3. The more resources, the less the relative cost
Scalability	1. Expensive 2. Time-consuming 3. Additional operating efforts	1. Lower price 2. Automated scaling 3. No operating resources required
Continuous integration and delivery (backup, testing, upgrades, maintenance, monitoring)	Manual	Automated

- *Software-as-a-Service* (SaaS) is the software which is working in the clouds. Typically, it means that the vendor provides an access to the application via the Web
- *Platform-as-a-Service* (PaaS) is a set of tools which eases the development and delivery of applications
- *Infrastructure-as-a-Service* (IaaS) is the environment provided for resource-intensive computing, storage and network management

As mentioned above, for many enterprises Hybrid Cloud is the most appropriate approach as it allows keeping some applications on-premises and integrating them with the cloud applications. The logic of refactoring the architecture may be the following—for those applications, which will not evolve, will not require large investments or have serious security issues, we choose private on-premises clouds. For those applications, which are delivered by vendors, SaaS may be used. Finally, PaaS and IaaS may be used for cloud enterprise application development and delivery (Fig. 3.7).

Continuous Delivery and Continuous Integration Currently, implementation of *Continuous Delivery* (CD) and *Continuous Integration* (CI) has become an integral

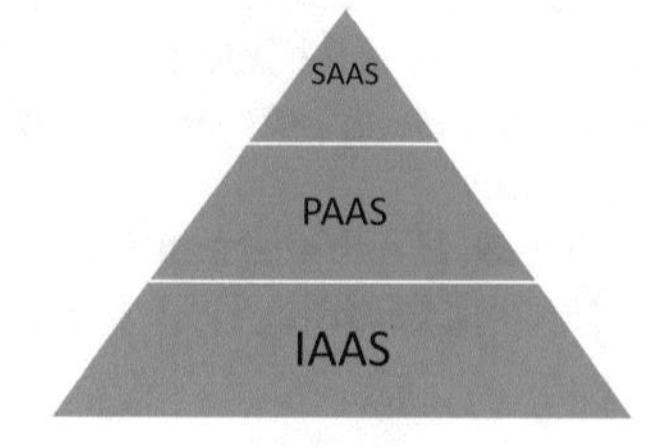

Fig. 3.6 SaaS, PaaS and IaaS

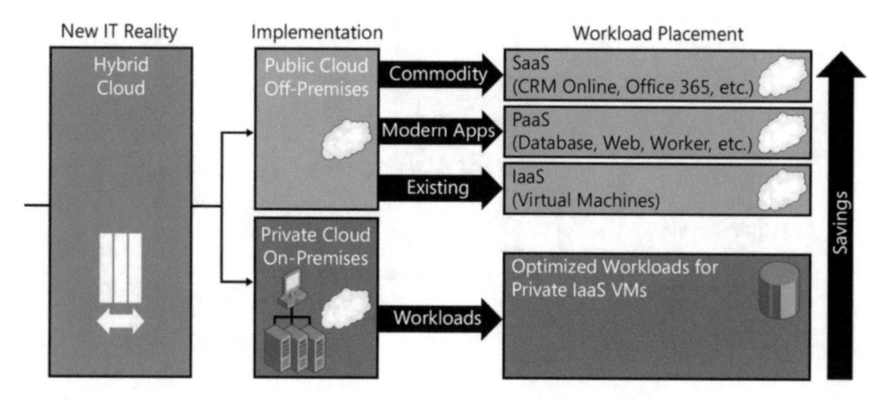

Fig. 3.7 Hybrid clouds

part of the enterprise software development lifecycle. Both principles underlie the Microservice approach, as far as continuous consistency and the availability of the system is very important. Despite the fact that the principles seem to be similar, there is a difference.

CI implies that all development changes should be integrated into the system as soon as possible. Each new code snippet integrates into the current version of the code, usually by making PUSH operation to the Version Source Control. After the code was published, the system should be automatically tested. There are various tools, which detect changes in the code and generate tests.

Subsequently, CI may be logically followed by CD, which provides safe and fast delivery of software into the environment. Frequent delivery promotes the agility of Microservices.

One famous tool, which supports both CI and CD principles is Jenkins. The main functionality of the tool is to execute the predefined steps (build, test), which are triggered by some event (or time).

DevOps The term *DevOps* implies the combination of development and maintenance operations and refers to roles and operations, which work at the confluence of development and operations. Such cooperation provides intensive collaboration of the different lifecycle stages and results in higher efficiency.

DevOps complies with Microservices approach. Microservices imply high correlation of all team players through all lifecycle stages. That is also very important for enterprise applications, as it allows continuous tracking of the change requests, an immediate response and implementing the required updates timely and coherently.

Despite the fact that Microservice approach does not have strict rules, it has distinct features. The analysis of these features and their implementation would allow us to understand the ways for automation and optimization.

The first key feature is componentization. A *component* is a software unit that could be independently replaced and upgraded. The logical decomposition of the system into components is a task for the business analyst and the system architect. The representation of the components via services is a task for the architect and the

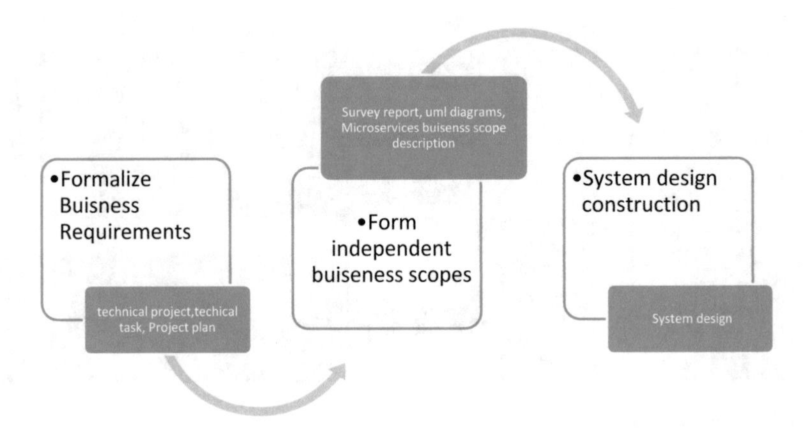

Fig. 3.8 Documentation flow

developer. These processes are not intended to be automated, but the junction points of the different actors' responsibilities could be processed by CASE tools.

Therefore, different team players participate on different levels of the Microservice system construction and implementation. Such participation could be logically and physically interconnected in a single process in such a way that the output of one level directly becomes the input of the other. The interconnection of these subprocesses is presented in Fig. 3.8.

API Gateway Pattern In general, the client service may wish to interact with all services in the system. As such, we need to support that interaction directly, which makes the system less scalable and more complex. A client would have to connect to all endpoints of each service. Besides, each service may use its own protocol, which could be not browser-friendly, and the client would have problems with it.

In order to avoid these problems, *API gateway* pattern may be used. API gateway is a single entry point into the system, which provides access to the system for all clients. API gateway may serve as a controlling unit, which performs authentication, load balancing, monitoring, request shaping, caching and static response handling (Fig. 3.9).

As API gateway encapsulates all services in the system and the number of the services may grow significantly, API gateway must be scalable. In order to decrease the complexity of API gateway implementation, specific libraries may be used, for instance, Spring Reactor based on the Reactor pattern [9].

Traditional asynchronous callback approach makes the application tangled, complex and vulnerable to errors. In order to avoid these problems, we may use a reactive approach. The most famous implementation is ReactiveX, which is supported by many platforms. ReactiveX extends Observer pattern and additionally allows for declarative composition of event sequences at a high level of abstraction. The Reactive Observable model abstracts from complicated callbacks and allows treating the event streams as easily as data streams.

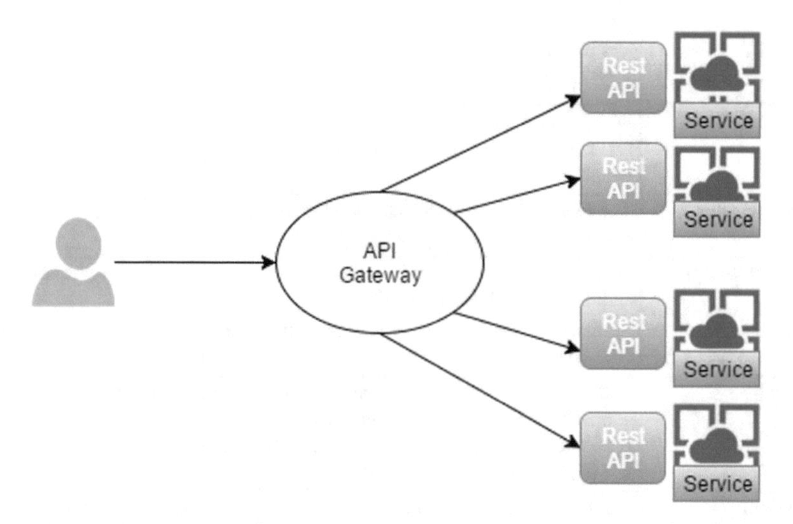

Fig. 3.9 API gateway

Circuit Breaker Pattern Another responsibility of the API gateway services is error handling. Without controlling the errors, the system may get blocked, while the functionality of the broken service may not be so crucial for the whole system. In order to avoid blocking, a circuit breaker pattern should be implemented [10].

A circuit breaker pattern stops cascading errors across services, monitors the system, provides fallback functions, and increases resiliency and agility.

For the circuit breaker we add a protected function call which monitors for failures. When an error occurs and reaches the circuit breaker threshold, the circuit breaker blocks the call of the function. Then, the circuit breaker monitors the state and keeps the function blocked until it recovers.

One of the efficient open source implementation of the circuit breaker pattern is Hystrix from Netflix.

Load Balancer Let us consider load balancers, which may implement various approaches to resource balancing. We will compare load balancers by Apache and Nginx (Table 3.2) [11].

Microservices Inter-Process Communication Patterns Naturally, Microservices arise the issue of inter-process integration. Let us consider the possible approaches and existing solutions.

There are two types of the inter-service communication: asynchronous and synchronous. The choice between these should be made depending on the system's needs. Let us consider both types, and present the most widespread patterns and technologies.

Usually, the Microservices are REST-based because of the protocol simplicity. The problem of using the pure REST communication is that it makes the Services communicate synchronously and directly.

Table 3.2 Load balancer policies

Policy	Description	Implementation
Round-Robin	Distributes requests evenly according to the server weights. By default servers have weight = 1	Nginx, Apache
Least-connected	The request is sent to the least loaded server in consideration with weights	Nginx
ip-hash(sticky)	Server is chosen according to the client IP address based on Hash function	Nginx, Apache Camel
Hash(sticky)	Similarly to ip-hash, Server is user-defined, and could be any key	Nginx, Apache Camel
Least time	Preferable Server defined by smallest number of connections lowest and average latency, which is counted either by header (time to first byte) or last byte (overall response time)	Nginx Plus
Random	The random server is chosen	Apache Camel
Topic	Requests are sent to all available servers	Apache Camel
Failover	If an error occurs on one endpoint, it tries another endpoint	Apache Camel

In synchronous communication, the sender waits for the service response. The most obvious problem lies in if the sender is blocked while waiting to the response. Potential failures should be addressed carefully.

By using pipelines with a REST protocol, we no longer need a direct connection between the services (Fig. 3.10). Now it is the pipeline itself that handles the messages. However, that makes services not self-sufficient, which is not a good practice for Microservices [12].

Asynchronous message-based communication is a well-known approach according to which the service places the messages into the channel, and one or more subscribers receive the message. If the sender only needs to inform the subscribers, the job is done. If it needs a response from the subscriber, the sender listens for the channel, but does not block its actions. There is a wide variety of open source messaging systems, the choice should be made based on the requirements and limitations of the services.

Usually, the services are not connected to each user directly, they follow an event-driven pattern, so that the services are subscribed to required events. For control purposes, a message broker pattern is used. Services place the messages in the queue and continue operation while the broker transfers the messages (Fig. 3.11).

Thereby, Microservices become autonomous. The problem is the complexity of the message broker. The flow of the messages becomes uncertain, and the possibility of the error increases.

There are several implementations provided by different vendors. The most common implementations of the message broker are Apache Kafka, RabbitMQ, ActiveMQ and Kestrel.

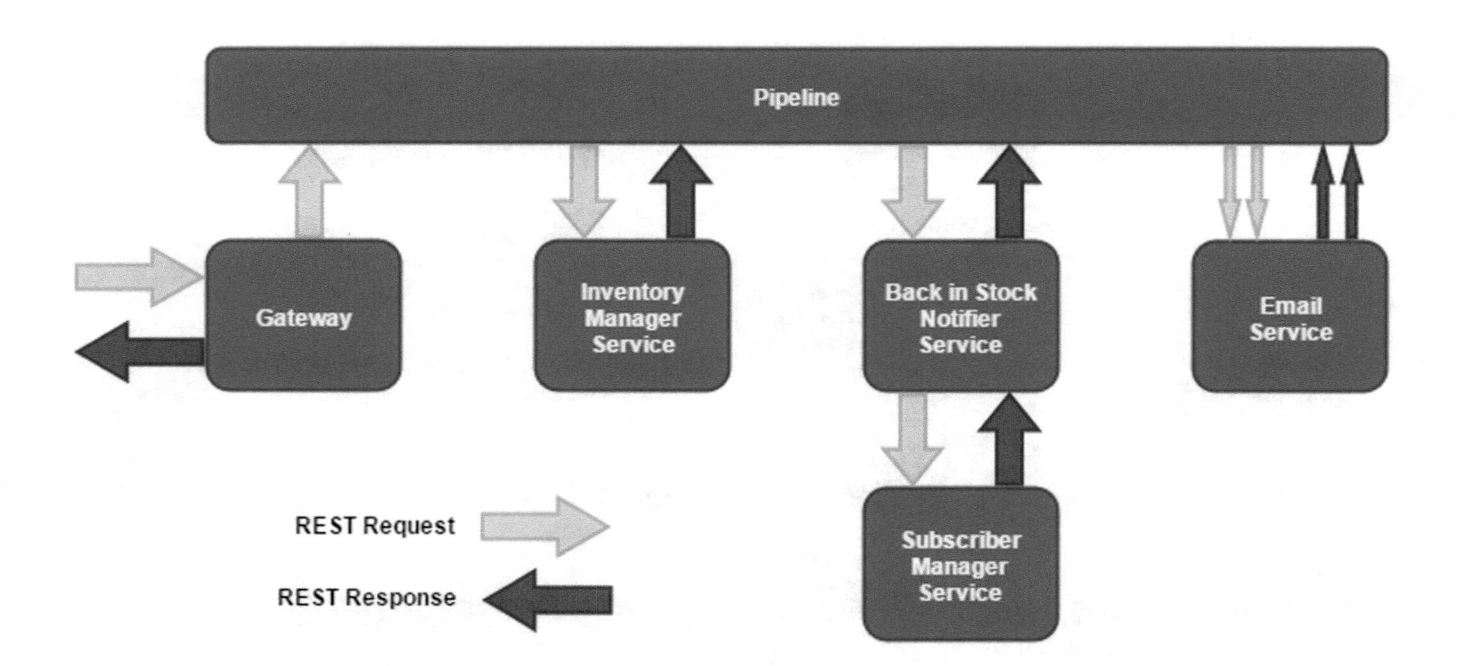

Fig. 3.10 Pipeline

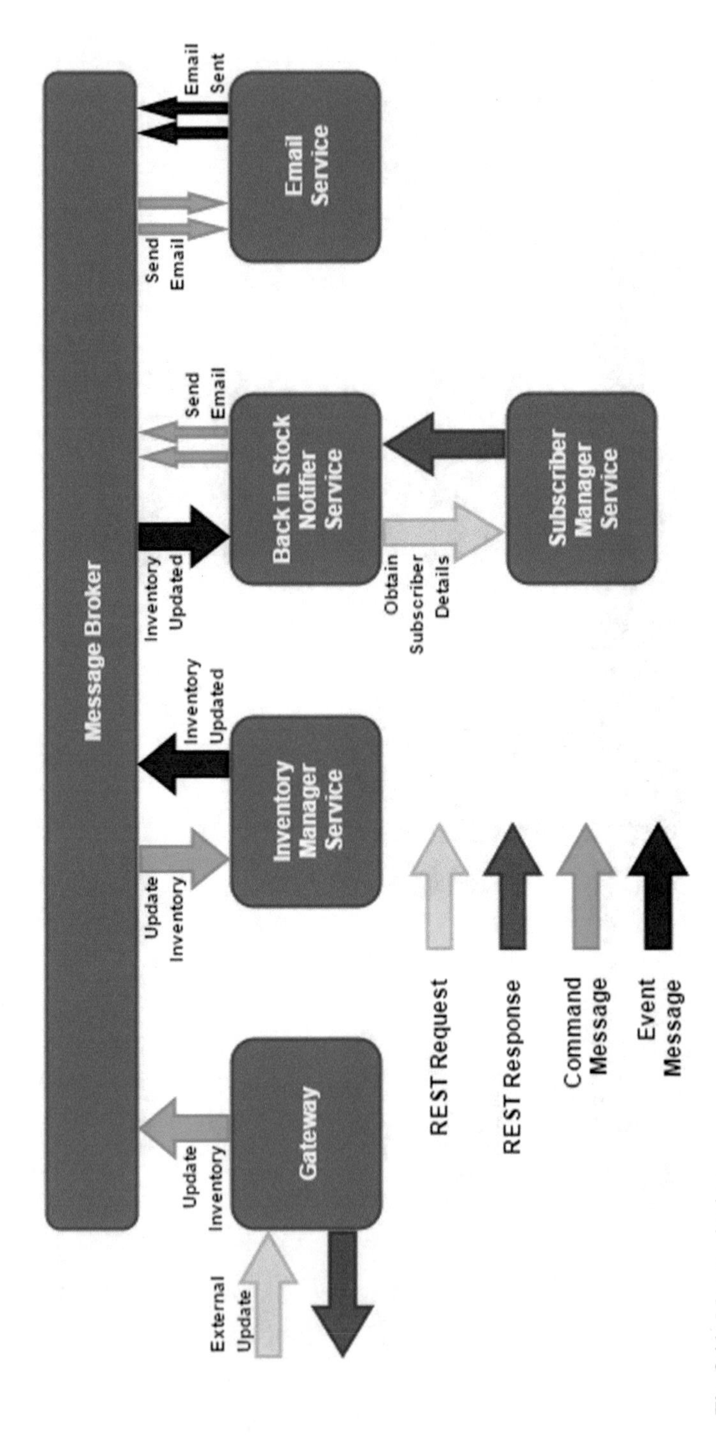

Fig. 3.11 Message broker

Therefore, the asynchronous approach with message broker best matches the highly scalable and loosely coupled nature of Microservices. Besides, asynchronous messaging with a message broker may be implemented by event-driven architecture.

Database Pattern According to the methodology, each Microservice should have its own database (Fig. 3.12).

Services may also share multiple schemas or tables of a single database. These two last approaches make applications less scalable, so the first approach is preferable. However, using multiple databases causes problems with data updates [13].

Event-Driven Pattern One of the solutions that helps to keep data in multiple databases consistent is event-driven architecture. Each time the data is changed in one database, the service publishes an event (Fig. 3.13). All subscribed services update the data (Fig. 3.14) [14].

In order to achieve the atomicity of the database operations and updates we can use an additional service, which acts as an event queue, so only local transactions are performed. Each service has access to the event table where it places the events. Event message service takes the first event from the table and publishes it to the message broker.

Another approach is to use a transaction log miner (Fig. 3.15) instead of an event table and event queue. After the service updates the database, it is reflected in the database transaction log. Therefore, the transaction log miner may read the log and publish the event to the message broker.

Another well-known event driven approach is event sourcing which implies that the event store (database) stores not only the current state of the entity, but also the sequence of the state-changing events. By replaying the events, it is possible to reconstruct the entities' state (Fig. 3.16). Event saving and publishing is an atomic operation; besides, all transactions make a well-structured log, which helps to track the operations easily [15].

However, pure event sourcing is insufficient if joint transactions of multiple entities are required. Let us consider the Command-Query-Responsibility- Separation pattern. As follows from the name, the pattern separates the queries and commands of the data (Fig. 3.17).

In compliance with the event sourcing approach, the list of events is stored in Write DB. When the command side publishes events after updating or aggregating the view, the query side consumes the event and recovers the DB state, which is stored in Read DB. After that it returns the data view to the UI.

Separation increases system performance and improves scalability. At the same time, the complexity of the system could be the reason for the errors, so the implementation should be very stable [15].

Service Discovery Pattern The network locations of the service instances change dynamically; thus, in order to refer to the service we need to use service discovery techniques such as client-side discovery and server-side discovery.

In client-side discovery, the client determines service instances network location and performs load balancing. That is performed via service registry.

In server-side discovery, services request load balancer, which is implemented apart from the client. After that, load balancer queries service registry.

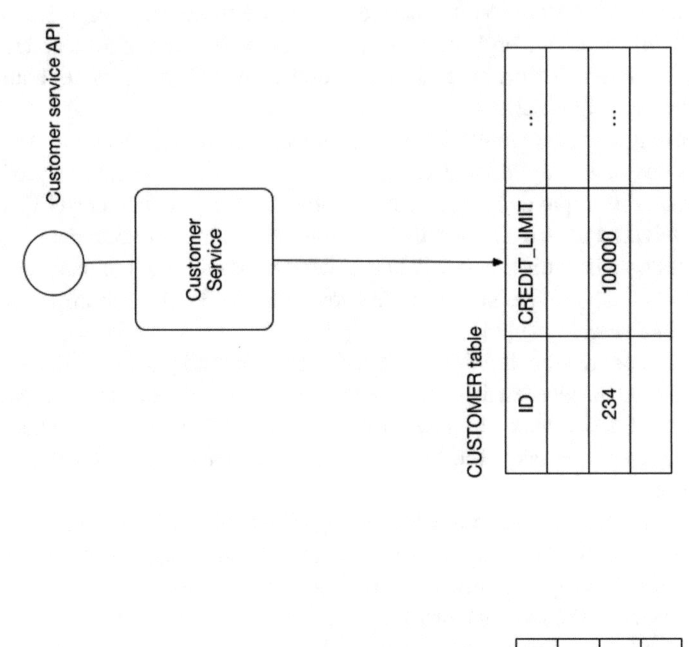

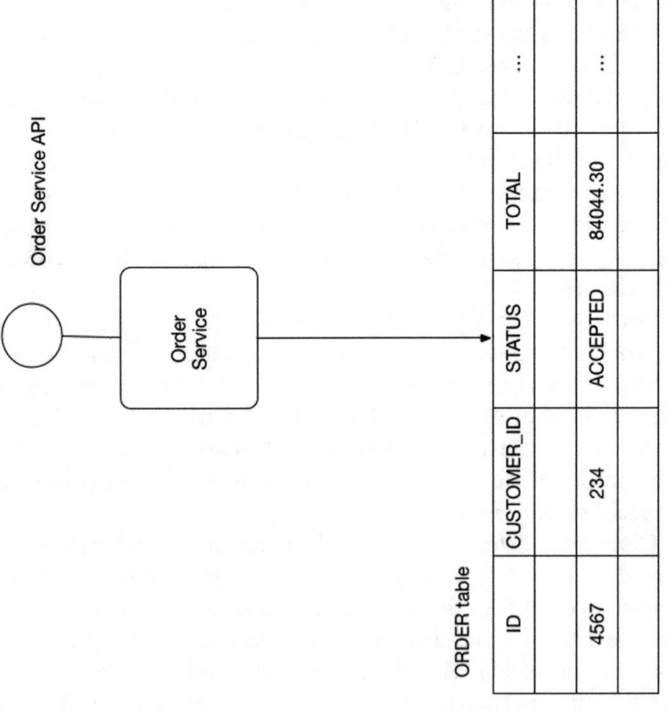

Fig. 3.12 Separate databases architecture

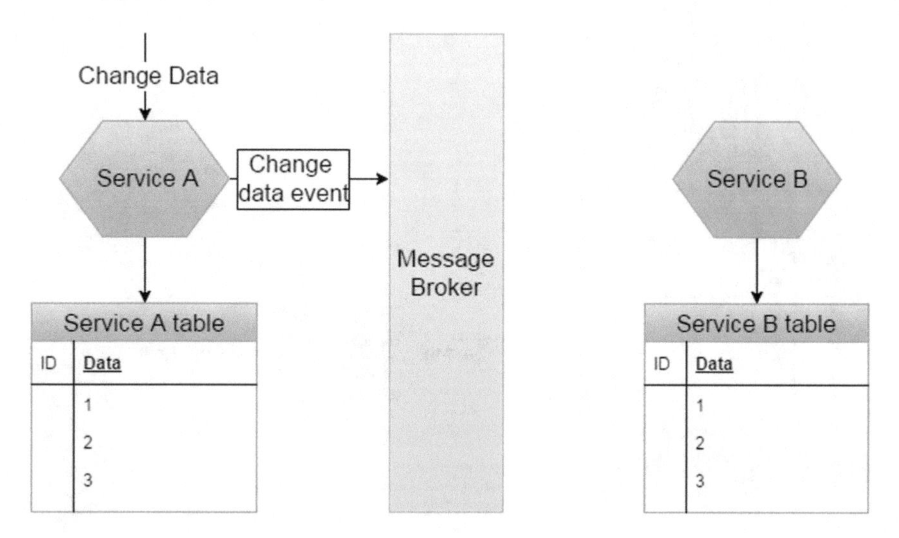

Fig. 3.13 Publish event

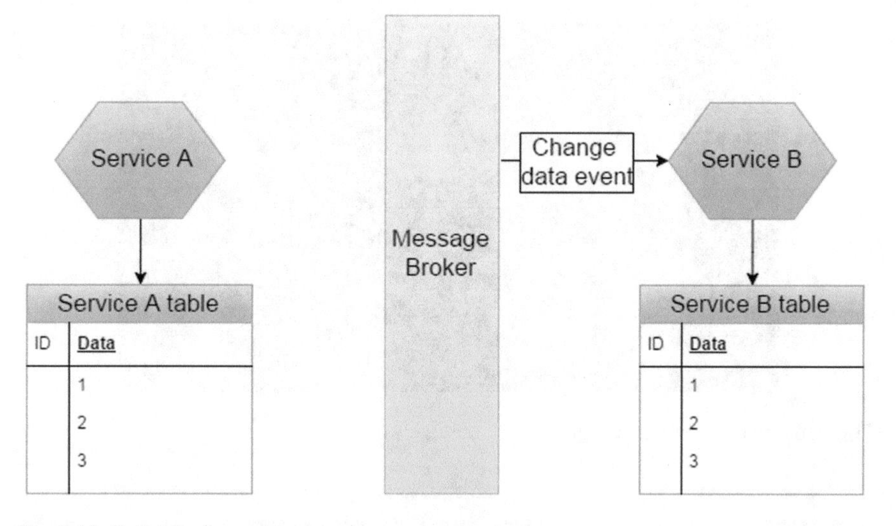

Fig. 3.14 Subscribe for event

Service registry is a database that stores service instance network locations. There are different approaches to service registration and deregistration.

According to the Self-Registration pattern, services themselves register and deregister the instances. The pattern is easy in implementation, but it requires implementing the registration function in each service. Another pattern is the Third-Party Registration pattern, which implies usage of the so-called service registrar that tracks services activity, when there is a new service instance it registers it with service registry. The

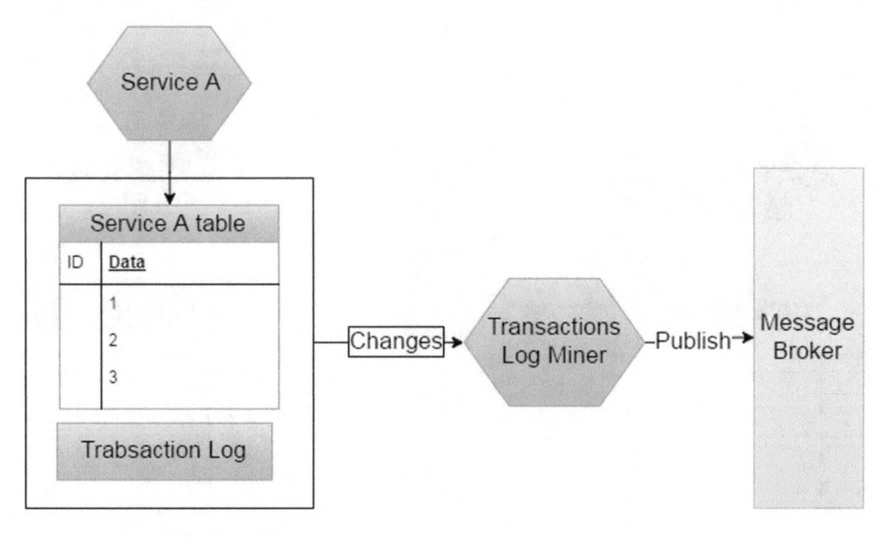

Fig. 3.15 Transaction log miner

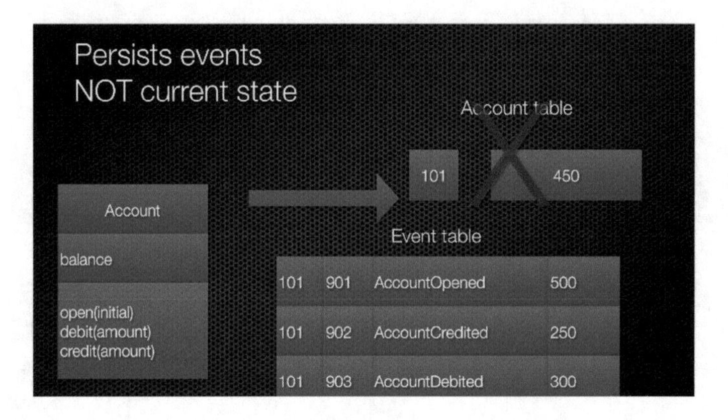

Fig. 3.16 Persistent events storage

pattern suggests a more complex structure, but avoids service registry functionality implementation on the client side.

3.4 Bank Microservices

The expansion of business and the scope of services provided by banks has led to an increase in the number of information systems within banks. In consequence, the integration problem arises. A bank has a need to integrate existing systems and new ones.

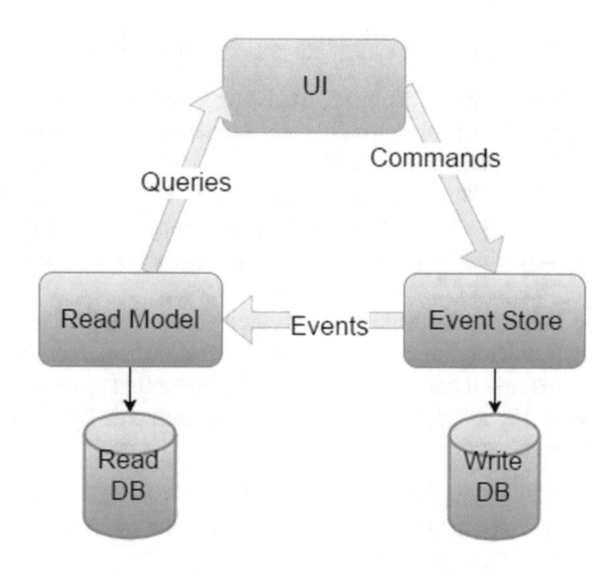

Fig. 3.17 Command query responsibility separation

Currently, banks have different information systems up their sleeve. It just happened for the automation of the Russian banking sector to be very fast growing and spontaneous. As a result, many banks have a large number of disparate IT systems that were developed and implemented without taking into account the requirements for their further integration into a single integrated information system. Moreover, some of them are already outdated and do not meet current business requirements.

Every year the number of new information systems in the banking industry increases, as well as the amount of services provided by banks. Therefore, there is a need to introduce new IT systems to banks in order to handle the load that is imposed by the growing services. The problem with this is that there should be a competent integration among those systems as well as between systems that the bank currently adopts.

Currently most of bank integration of IT systems is done point-to-point. This means that every time when a new system is implemented, it is required to redesign and reimplement the integration layer as there is a large number of interconnections. This means an increase in costs both to implement it and to support it in the future.

The need for integration arises when multiple information systems are used in an enterprise, each of which addresses a different set of business objectives or different parts of the business process flows on the side of the remote offices or partner companies. Adding a new application to the system or replacing one of the applications requires the establishment of relations with each of the existing applications. One of the most effective ways to solve this problem is to use a central integration platform based on SOA [16]. An integration platform that acts as a bind layer enables a business to provide its customers with all the services of the enterprise while hiding their implementation. This simplifies and accelerates the introduction of new components, reduces risks and makes the information system more manageable and agile.

The benefits of using integration platforms [17]:

- Preserving investments in existing systems by creating a single transport line based on a universal messaging system between applications
- Increasing efficiency of business management through the use of consistent information from multiple systems and the integrated organization of "intelligent" data exchange between different applications in a unified form in real time
- Reducing the impact of human factors and the number of errors during the manual transfer of data from system to system
- Reducing the time spent manually searching for information in different systems
- Reducing the cost of system integration, which will be created in the future, by making it easier to connect to the integrated system of applications and databases
- Quick responding to business requirements by means of flexible user-oriented services.

The architecture for the integration of bank systems is based on a particular case study. In future, this architecture can be used for other banks taking account their specific features.

This case study refers to a problem that arose in a large commercial bank (from Russian TOP–40), which has an extensive branch network in the Russian regions, and a head office in Moscow. Each branch of the bank has a separate installation of the automated banking system (ABS) from a K1 company. To service customers in the bank's offices the company uses ABS interface. The current accounting system is decentralized—data between all of the individual ABS installations are synchronized after the end of each business day for each of the branches (Chap. 1 gave another case study on Dapps). There is at least one day difference between the actual data and the general ledger data.

The bank plans to improve customer service and increase its share of the credit and card service market and at the same time reduce labor costs and improve the image and reputation. Its strategy is focused on the development of the bank, and the growth of its key financial indicators. To achieve these goals the bank's board made a strategic decision to improve the IT infrastructure through the implementation of the following information systems:

- Front-office (with credit and card conveyor support)
- Centralized automated banking system (CABS)
- E-banking system for individuals

At first stage of the project the bank required the first two systems. After the end of the first stage, the bank required the e-banking system.

According to the results of the tender the bank decided to conclude a contract for the implementation of the front office and the CABS with K2 company, and for the other two IT systems, the bank chose K3 company. The most labor-intensive and time-consuming was the introduction of CABS, since it could not run as a monolith, but needed gradual expansion after the introduction of each branch.

As a result, the problem can be formulated as follows: for the commissioning of the new integrated IT infrastructure it is required to integrate the implemented IT

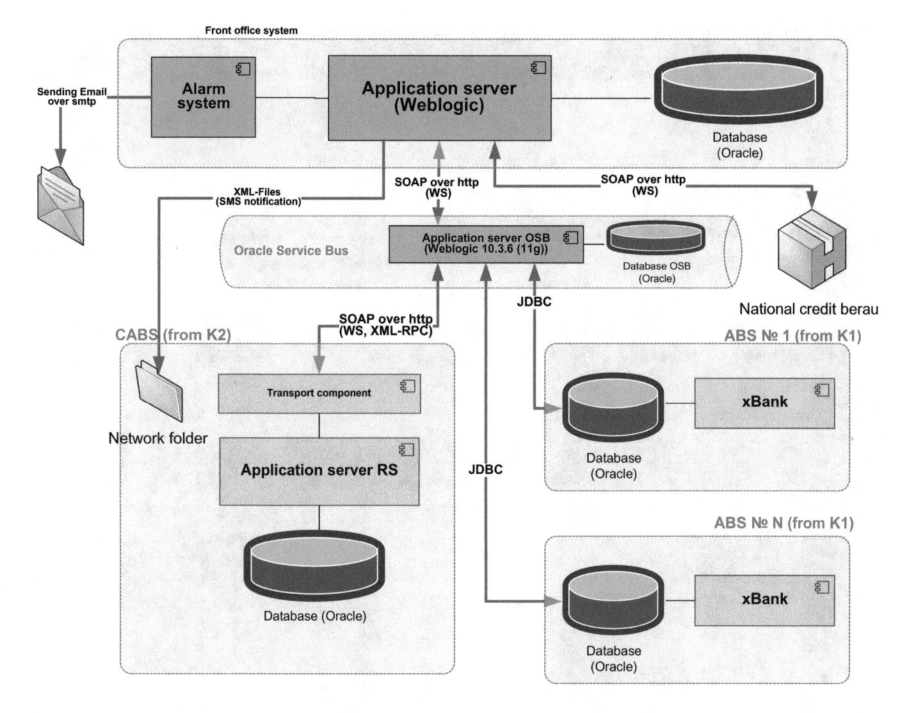

Fig. 3.18 Prospective bank's IT architecture

systems and the existing IT infrastructure into a single information system within the bank.

The bank's prospective IT architecture is shown in the Fig. 3.18. This new IT infrastructure consists of:

- Integration application—transforms and transfers requests between the front office and various bank's ABSs;
- Adapter to the front office application server—receives requests from the customers, and sends them to the ABS of the Customer; sends status reports on request implementation/processing back to the front office;
- Adapter to CABS (from K2)—transfers appropriate requests from front office to CABS for transactions on customer contracts, sends implementation/query status reports;
- Adapter to ABS (from K1)—transfers appropriate requests on front office operations and customer contracts to ABS, sends implementation/query status reports.

The interaction between adapters within interfaces is carried out in the synchronous mode.

The modules operate in multi-threaded mode, so different requests are processed in parallel.

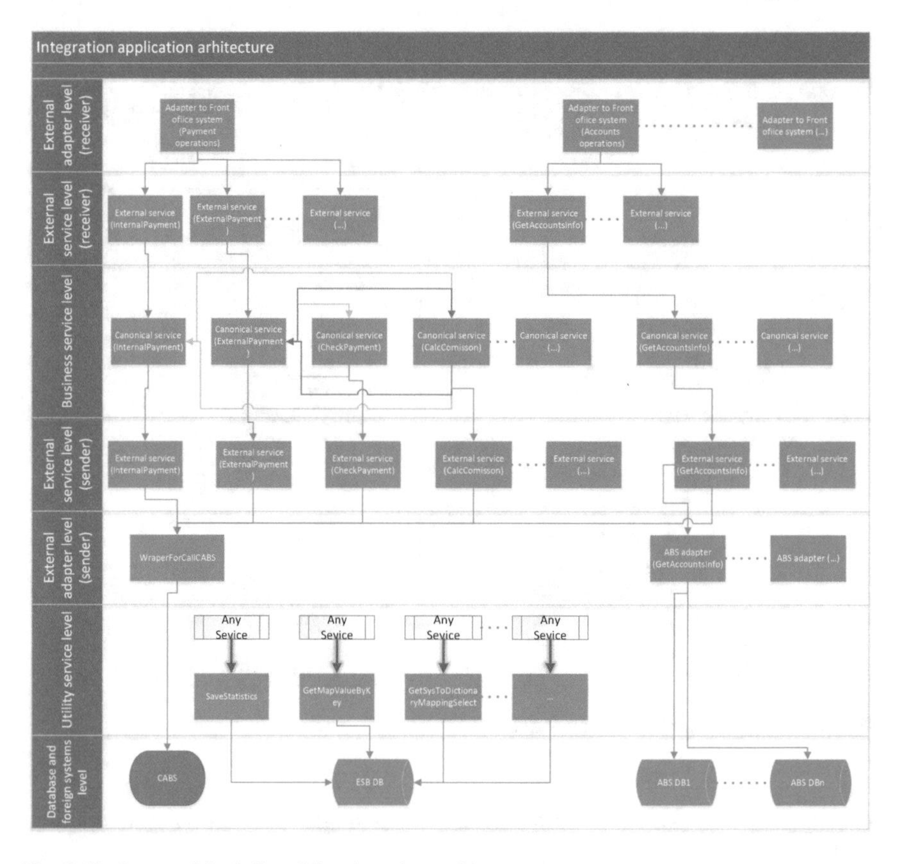

Fig. 3.19 Proposed "to be" multilevel service architecture

The architecture is shown in the Fig. 3.19, each level of the architecture is described below.

External adapter level (request) is the layer of services (adapters), responsible for interaction with the calling system (e.g., the front office, online client for individuals, mobile banking, etc.). This layer unpacks messages from the format described in the WSDL service scheme, and packages messages back when answering the calling system.

External service level (request) is the layer of services responsible for the message format transformation to the standard format.

Business service level is the layer of services, in which the basic business logic resides. Each individual service performs a specific function. This level of service can invoke the services database level, and any number of standard services, if necessary (for example, to implement complex business logic). This layer can be synchronous or asynchronous.

External service level (response) is the layer of services, responsible for the transformation of the standard message format to the format of the called system (ABS, CABS, etc.).

External adapter level (response) is the layer of services (adapters), responsible for packaging the message in conformity with the format of the called system (e.g. SOAP, XML-RPC, JDBC, etc.).

Utility service level is the layer of services, responsible for interaction with the database and system settings.

Database and foreign systems level is the Oracle database, which stores the tables for ligaments, dictionaries, system settings and CABS called via XML-RPC.

Because the functionality of the system was designed as a Web service, there is a need to facilitate the further development and modernization of the system. This is so that if the company later decides to implement other access interfaces, or to add a new module to the system (for example, a subsystem for remote banking services), it will be enough to develop the adapter that will be deployed on the bus standard service, abstracting from the details of its implementation. This architecture can be customized for any other bank.

The implementation of this integration pattern will improve business processes of customer service and document management. The bank will provide rapid deployment of sales outlets, save time on the of new product development, and improve agility.

3.5 CRM Microservices

Nowadays there are a lot of companies that build their business around clients and deal with them directly. Therefore, a lot of Customer Relationship Management (CRM) systems appear, which focus on satisfying customers to increase company profits in the long-term. CRM systems provide wide range of functions: logging deals, providing leads, preparing reports and more. Using these features each employee of the specific company can work more efficiently and focus on the final goal—the sales deal. To achieve these target analytics, management and directors analyze information which comes from CRM to plan their next steps.

There is where *geo-marketing* comes in. Geo-marketing is a relatively new term in marketing and economic management. It is a new concept where "location" plays main role. It is used for a wide range of problems, such as: planning process, selling products, managing spatially-distributed objects that describe the infrastructure of the territory, consumers and the competitive situation. A common usage of geo-marketing is geo-marketing research. Geo-marketing research is collecting internal and external business data which are connected with geography, analysis of the data and reporting on it. Reports can relate to the current state of the key indicators of the company, provide new places to focus on (e.g., open point of delivery of goods). This process requires huge amounts of resources. That is why companies try to automate each particular research and develop systems which provide all the steps of analysis with updated datasets. Nevertheless, such systems are built with ready sets of data provided by researches, with one particular goal (e.g. maximum efficiency of outdoor advertisement), and usually do not connect with any existing enterprise systems.

Combining the concept of geo-marketing and the fact that CRM systems are widespread we get the idea to develop a geo-marketing system integrated into CRM. Some researchers say that about 70% of all business data in CRM contain a geo component such as address or coordinates. Such geo-marketing systems can automatically collect data from CRM and using external data, such as Public Cadastral Maps [18] or Open Data Portal [19], can provide geographical analysis and reports.

Geo-marketing usually deals with the following problems:

1. **Territorial planning at the macro level**: The main goal is to select, with help of geo-marketing data, territorial entities within the borders of the state with the best potential for business. For this, the system should contain enough quantitative information about each particular part of the state
2. **Territorial planning at the micro level**: This problem deals with detecting the factors which influence passenger and car flows in particular areas of settlement (e.g. town, village, etc.)
3. **Socio-demographic analysis**: The result of solving such a problem is a geo-demographic map which consists of layers of the population's demographic density, depending on their movements between zones and geographical objects, and their common parameters (such as sex, age, income level, number of family members etc.)
4. **Direct marketing**: Geo-marketing identifies the location of the target group of consumers. Further, communication technologies support efficient bidirectional communication
5. **Market analysis**: This approach allows building of a multi-level dependency on competitors' pricing policy, customer satisfaction and preferences, communication activity, distance from competitors, density of consumers etc.
6. **Analysis of location**: There are two inverse problems; first is identification of the "best" location for a new trade point based on a set of parameters. Second is identification of thes parameters (e.g. types of goods to trade, opening hours) for a particular location
7. **Advertisement and media planning**: Determines the size, direction and traffic of the customer flow, and manages targeting
8. **Risk analysis**: Clarifies the potential problems that may appear in case of deviations

Companies seldom address some of the above problems (e.g. territorial planning at the macro level), whereas they have to respond the others more often. A geo-marketing system should not include methods for solving each of them. There is a number of companies who provide specific geo-marketing analysis for global problems like territorial planning.

3.5.1 Analysis of Existing Solutions

The leading companies who provide geo-marketing research are:

- "One by One Consulting Agency"
  This company solves the location analysis problem by providing access to their cloud system with a large amount of open data. They have no integration tool for their customers. They also provide potential profit reports for a chosen location
- "Smartloc Geo-Marketing Company"
  This company provides a wide range of geo-marketing research to retailers, construction, and administration. Research is targeted to a particular problem. There are companies who provide the same services, but those two are more widely known than the others
- "GFK"
  Provides mostly the same services as previous companies, and geo data for sale
- "Data+"
  Founded in 1992 as a joint venture between the Institute of Geography of the Russian Academy of Sciences (Moscow, Russia) and Esri (Environmental Systems Research Institute, Inc., Redlands, California, USA). The company provides a wide range of services (e.g. geo information data) and makes geo-informational system based on ArcGis technology stack. They have built geo-marketing systems for solving specific problems (mostly for location analysis) for companies like Sberbank or X5 Retail Group
- "Geocenter consulting"
  This company has mostly same range of services as Data+, but recently most projects have been connected with selling geographical data. The main difference between the services of this company and their system (Betelgeuse) is: Betelgeuse will provide integration between existing customers' CRM systems and will not have any additional features. Also, it provides different types of geo-marketing features and different types of reports. Third type is CRM systems with location intelligence module
- "Galigeo"
  Galigeo has stand-alone software for geo-marketing and location intelligence, but also has an integrated tool for SalesForce CRM to provide location intelligence and geo-marketing features with data usage contained in SalesForce. This software is relatively new and a direct competitor to Betelgeuse. The difference between them is that Betelgeuse will integrate not only with SalesForce, but also with some other applications.

The common features of Galigeo for Salesforce are: visualization and adding addresses to leads, customers etc. These features identify the entities on the map, and visualize their key economic indicators and socio-demographic data; they indicate the zones, where the company has low sales or no sales at all; the sales/visits statistics are integrated with the Salesforce calendar [20].

3.6 Cloud Services

3.6.1 Process Modeling for Virtual Machines in Clouds

Currently, the use of virtual machines (VM) has increased significantly for cloud-based service infrastructure [21]. Using a layered architecture and technical infrastructure (TI) simplifies the deployment tasks, setup, transfer and support. In addition, the use of VM reduces the dependence on the setting of a specific configuration of physical servers.

One of the main problems with the use of VM is the forced decline in performance, compared with classical architecture-based TI [22]. This problem is caused by the forced separation of server resources between multiple VMs. The problem of performance degradation when using VM, as well as a significant increase in TI administrative complexity [23] is one of the most vital. This challenge is also confirmed by the increasing number of cloud platforms [20], in which the end user gets only a set of VMs allocated for these resources, but does not have access to the real servers. The problem is further complicated in the case of a heterogeneous technical infrastructure, in which different equipment is used for a variety of computing tasks.

For solving the problem of the optimal configuration of VMs based on dynamically changing external conditions (system load, TI availability etc.) we propose a methodology of VM auto-configuration. The proposed methodology involves creation of a central domain control agent [6] followed by its reinforcement learning. Such an agent allows for each state of the system to configure all of the VMs so as to optimize their performance, taking into account the available allocated resources. This approach solves the following problems: it allows taking into account the external changes, increases the system response, reduces the influence of the human factor. However, there are a number of constraints for its application. One of them is its low scalability for productive use. A large number of VM configurable parameters, together with a large number of VM themselves generate a tremendous number of parameters of the function, which can only be optimized for a time exceeding the lowest possible productive system performance or greater than the time of change of external conditions.

To increase the scalability of the auto-configuration agent training algorithm we propose using of models of information processes performed by the system. For the modeling of information processes we use the applicative approach to modeling processes—π-calculus with types [24]. As part of this approach, a model of the process is a term containing the description of the data transmission and reception. In the simulation of information processes performed by a set of virtual machines, each system function, the execution of which is significant for the overall performance, is associated with an elementary process of receiving the initial information and the produced result. The scope of the computation involves channel names which define access to functions, and information on the operation is coded in the channel type and depends on the channel type and the types of data being transferred. We chose the applicative approach [25] due to the following reasons:

- No fixed structure of the process—the local behavior of certain subprocesses can be determined without changing the entire model
- No difference between the functions and data—depending on the context of the application and the name of each context, π-calculus can act both as a data block, and a channel of transmission/reception, which simplifies modeling operations on metadata and service operations
- Direct computability—any valid π-calculus term is an executable program for the corresponding abstract machine, so there is no need for additional compilers, the model is modifiable and agile
- Multi-level type system—static typing is the key to behavior analysis of the information system, and an elaborate type system significantly increases the power of the model.

The enhanced auto-configuration algorithm reduces the number of configurable parameters due to indirect configuration. The configurable parameters of VM are encapsulated by applicative properties of the environment in which processes are performed—its types that are assigned to channels transmit/receive data. Each type corresponds to the set of the VM settings, which is significant for some of the functions and influences the metrics. Types may be assigned either manually or using type constructors. Assigning the type of reception or transmission channel is interpreted as the operation in certain VM settings. The valid model is such that the settings obtained from each type are applicable and consistent. Due to the fact that the performance computation is based on the types of the model, a valid assignment of the names to all type variables provides an estimate of the overall performance.

The following outline of algorithm is proposed. Based on the actions performed the process model receives the current load, and using pre-defined types or adapting them to the level of detail, the algorithm performs an initial assignment of the correct type in the model. Next, we apply the Q-learning algorithm, which is based on the collection of samples of remuneration and evaluates these types. The algorithm gets the correct assignment types, maximizing system performance. The algorithm gets the configuration of all VM corresponding to the correct assignment types. The architecture is shown in Fig. 3.20.

The proposed approach combines applicative modeling with reinforcement learning, and allows the use of previously developed methodology to automatically configure a large number of VMs with shared resources, taking into account the increasing number of configurable parameters.

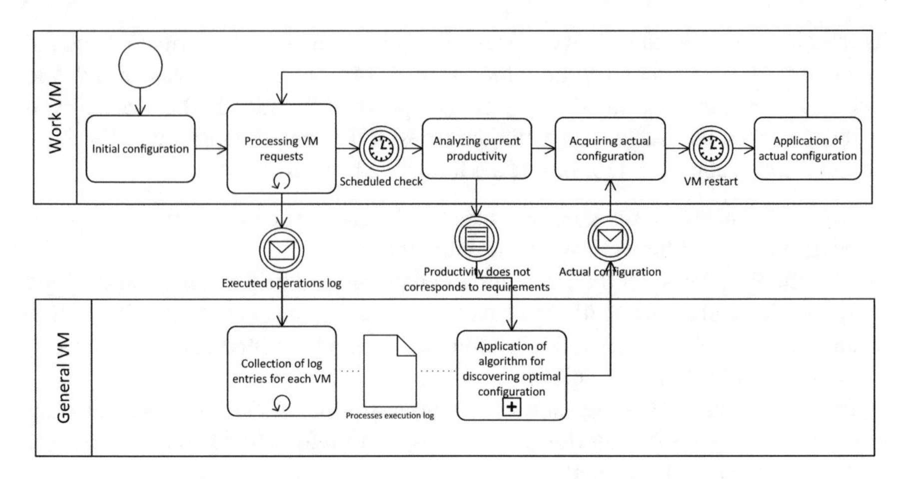

Fig. 3.20 VM auto configurator architecture

3.6.2 Information Process Model

We use an applicative approach—π-calculus for modeling information processes executed by a VM cluster. The process, according to this approach, is a correctly formed calculus term. The model of process is the reduction of such a term with an abstract machine [26]. In the context of information process modeling, reduction applies not only to the term, but also to the context which contains a description of process environment. A process is executable if there exists such a configuration of the abstract machine, such an environment context and such start conditions, that the term could be reduced to an elementary empty process.

Further, we present the grammar for the extension of a π-calculus type system and an abstract machine for modeling and analysis of information processes in several VMs. We apply the developed model to acquire input data for Q-algorithm, which optimizes the virtual machine configuration.

Classic π-calculus uses the following rules for building process terms:

$$P = a(x).P \,|\bar{a}x.P|\, \nu a.P|P|Q|\mathbf{0}|!\,P$$

With regard to the problem of modeling the processes performed by a cluster of VMs, these rules are treated as follows:

The data reception and transmission: $P = a(x).P|\bar{a}x.P$. The data reception prefix connects the name x in process P; the data transmission prefix does not bind any names. Combinations of these prefixes describe the sequence of receiving input information for the launch of the information process, its internal routing, processing, and returning the result. As π-calculus has no syntactic differences between the data and channels, these operations allow modeling processes with dynamically changing structure, when the result of function is a new data handler. This feature allows simulating complex interfaces, factories and a variety of processors in a VM.

Restriction: $P = \nu x.P$. This operation binds variable x in process P. A variable related this way is not defined by the substitution rule. The information process model built for multiple VMs performs various functions. Restricting certain channels or receiving data simulates the performance of functions within a single VM. Restricting input data blocks simulates sending them only to specific processors. Unbound channels can interact with any of the VM data, unrelated data blocks may be processed by the VM.

Replication: $=!\,P$. The application of this operation creates the process for which during the reduction it is possible to obtain an independent process, carried out in parallel. This operation is necessary to simulate the background sub processes, which are initiated each time processing relevant information is required, possibly several times during one iteration of the current process.

Parallel processes: $P = P|Q$. This operation is the main method of constructing new calculus terms. Generally α-reduction is applied for two processes combined in parallel execution. Parallel processing allows modeling all the flows of information that are performed at the same time, interacting and sharing the results.

Reduction of terms of π -calculus is carried out by the abstract machine according to the following basic rules:

1. $\mathbf{0} \rightarrow \mathbf{0}$
2. $!\,P \rightarrow P|!\,P$
3. $P \rightarrow \nu x.P\, if\, x \notin fn(P)$
4. $\nu x.(P|Q) \rightarrow \nu x.P|\nu x.Q$
5. $\bar{a}x.P\,|a(y).Q \rightarrow P|\,Q[y/x]$

The main metrics of the implementation process are supported by variables controlled by the abstract machine. These meta-variables that do not relate directly to a running process allow a comparison of different processes. Types of variables and allowable transaction processing are not described within the process modeling system. Interaction with these variables cannot be performed by the functions of the process, but only by the meta-function—abstract machine operations or functions, processing execution contexts. These variables will be used to store the indicators to assess the performance of the selected VM configurations and the VM set will be used in the algorithms to find the optimal configuration.

To increase the expressive power of the models, we use typing [27]. The essence of this is to assign a type to each named computation entity and to extend the abstract machine by the rules for type-checking.

The type of a name can be simple: $x : \alpha$, or it can be a type of data transmitting channel x $:\uparrow \alpha$, or a data receiving channel $x :\downarrow \alpha$. Thus, any correctly typed elementary data receiving process must have the following structure:

$$\bar{a} :\uparrow \alpha.x : \alpha.0 \,|\, a :\downarrow \alpha\,(y)\,.y.0$$

Typing can further validate the constructed terms. The simulation of information processes assigned to the cluster types allows a further description of the function, simulating different subprocesses. In addition, typing is used to simulate the functions of the consumed VM resources. To emphasize that the type name is also a feature of the run-time, not only the design time, assigning a specific name, and type of computation will be reflected in the context of the implementation process. Context of the evaluation will be called as a set of pairs of: "$name \rightarrow type$".

When using the execution context, the first step of the process is the initial filling of the context by pairs of values defined by the process model. Further, during the reduction for the operation, which requires type checking, for each name a corresponding value to the projection of the type of the execution context is obtained. With the application in parallel of alpha conversion for the name in the term, the name is replaced in the type context. This approach is necessary because the same name may be used for a function of the data channel transmission/reception unit or the information processed.

The resources consumed by processes running within the VM, are modeled as follows:

- Algorithmic resources, such as the maximum number of simultaneous requests processed [28], are modeled by introducing the technical processes in the configuration of the abstract machine. The algorithmic resources of each VM are modeled including a subprocess term for the data transfer to an external process and obtaining processing results. As an external process is used, a subprocess term is added when performing replication. Such technical process models the restriction on the maximum number of copies. Such an operation is written as follows:

$$!\,P^\wedge n = P|!\,P^\wedge(n-1)$$

For technical resources, the algorithmic process should have a structure $a\,(x)\,.\bar{b}x.0$, where the type of the data receiving channel should consider the time spent performing the subprocess depending on the type of the received data block x, and channel type b, and must consider the costs of return to the main process. Therewith:

- Hardware resources are modeled with the help of the types of names assigned to calculus
- The time required to perform the functions of the CPU [29] (measured in operations per second) relates to the types of assignable channels that transmit/receive data
- The block size of data (measured in bytes) is stored in memory [30] with the assigned name(s) that at this stage of reduction serve as argument(s)

Thus, when changing the role of a specific name, various projection types are used. To meet these requirements within the context of the implementation of the process types, we use the following assumption: each type is an object that contains information about possible operations applicable to it, including the static information, which does not depend on the role. The function gets the type of the execution context, takes as input the name and role, returns the current projection type and writes the abstract machine variable changes, as well as the changes determined by its static components.

3.6.3 Optimization of Virtual Machine Configuration

Optimizing algorithms for the configuration of VM, cluster parameters on the basis of the information process models must be implemented on the basis of the following assumptions. First, the algorithm should be suitable for machine learning. Due to the auto adjustment of the VM configuration in response to environmental changes, the algorithm should be able to run without human intervention. Secondly, the algorithm should produce a VM cluster configuration based on the account of running process update. Any VM configuration update requires a complete reboot of the virtual machine. This is a technical limitation of the algorithm. Third, the algorithm must be adjusted to find the optimum value for a cluster of VMs.

We propose the following optimizing algorithm:

1. Cluster processes are based on the server logs for the period since the last run of the algorithm. The model is automated and may be adjusted manually, if necessary. The initial model of information processes includes functions and data types
2. Initial configuration of the abstract machine includes:

   - Data input transmission process

     $$!\, i_1^{n_1}(x_1).!\, i_2^{n_2}(x_2)\ldots !\, i_m^{n_m}(x_m).$$

     Each element $!\, i_m^{n_m}(x_m)$ is recorded in the data logs of the server. Parameters $N_1, \ldots, N_m$ describe the current load and the number of input parameters
   - Initial configuration parameters have the weights that indicate memory usage for data and CPU utilization; they are based on server logs

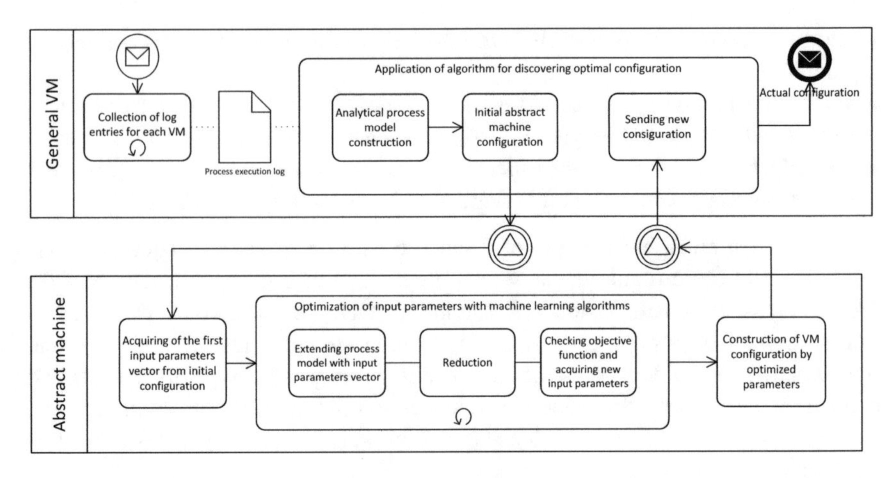

Fig. 3.21 Algorithm of VM optimal configuration building

- Initial configuration variables. For each type of resource optimized, the abstract machine adds one variable of the appropriate type with a value of maximum resource. The selected configuration variables refer to the parameters of VMs

3. Reduction is useful if:
   - The information process with the input data is reduced to an empty process 0. This reduction shows that the model is correct and suitable for further analysis, and that the suitable input data set correctly simulates the cluster of VMs
   - None of the variables of the abstract machine can take a negative value
4. Input parameter optimization uses machine learning algorithms. The input of the algorithm is applied to the volume process input data vector and VM resource configuration. The output of the algorithm is a vector of residuals of the abstract machine code, after the complete reduction of the term. The task is to select the minimum configuration vector and the maximum input vector.

The algorithm is shown in Fig. 3.21.

The applicative approach to modeling auto-configuration of VM can increase efficiency of the known methods of evaluating, configuring and testing distributed systems and cloud services. In particular using its output as an input to the Q- algorithm significantly increases the number of parameters to be optimized by combining them in the calculus terms.

Another advantage of this approach is obtained by transactional testing methods such as TPC-C. The advantage is achieved by applying trained models to the assessment of the configuration and further implementing this configuration in a cluster or cloud server. This double application improves agility. The approach requires minimal technical expert involvement and often operates unassisted.

We propose a method for automated VM machine configuration under a changing load. This method uses machine learning algorithms with reinforcement and models of information processes. The models include π-calculus. The optimal configuration involves Q-algorithm which reduces the π-terms by means of an abstract machine. The proposed method uses an applicative approach to model the information processes executed by VMs.

Due to high flexibility, i.e. agility of the type system, the method supports heterogeneous types of VM resources including algorithm-based access to the central scheduler and hardware resources such as memory, CPU usage and power consumption. The method imposes no extra constraints on the internal structure of VM clusters. The proposed VM mathematical and methodological concepts are essential for the implementation of language structural modeling and can be executed directly by the abstract machine. This allows for agile analysis and adjustment of the VM configuration.

3.6.4 Experimental Design of Automatic Virtual Machine Configuration

The need to balance resources between multiple VMs running on the same physical facilities has been noted in numerous studies [31–33]. Often, in order to achieve resource balancing, methods based on the interaction between the main machine (host) and the working VMs are proposed. These methods include content-based sharing, ballooning, memory compression, and page replacement [34, 35].

Each of these methods allows a limited number of parameters for each resource: e.g. maximum expected load (peak), average VM load and current consumption of the resource. As mentioned in [36], in order to reach the best performance of resource redistribution for the VM cluster the data is required for expected load for a certain planning horizon of resource allocation. Another significant limitation for the scheduler is the specificity of operations. This is mission-critical for grid [37] solutions, where the computational environment is heterogeneous. In this environment, the performance of the entire system can increase significantly if the most suitable VM resources are selected.

To achieve these objectives, we suggest the applicative approach to the model VM information process by means of the π-calculus [38] for each individual VM. This section proposes an architecture that implements this approach in a test environment. The proposed solution includes a tool for logging, a resource scheduler, and a static resource scheduler/optimizer for each VM.

The architecture is shown in Fig. 3.22.

The VM functions with the switched-on logger. Upon being post-processed, the logs are transferred to the input of the analyzer. The analyzer yield is the model of the VM information process. The model of the process is a valid term of π-calculus. Based on this model, the configurator composes the static parameters for each VM, and the schedule. The schedule is bound to the process steps. Based on the analysis, the configurations are applied to all VMs and the schedule is passed to the scheduler.

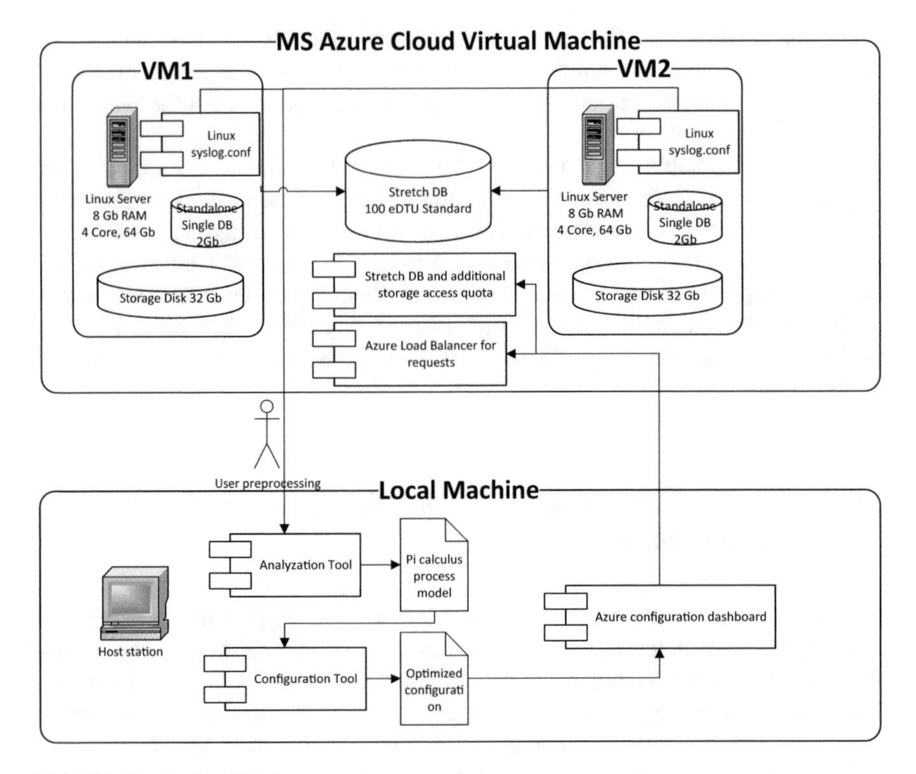

Fig. 3.22 Test bed architecture

The scheduler communicates with the logger through the analyzer which signals the execution of each process step. Depending on the duration of each step and the resources required by every next step of the process, the scheduler redistributes resources of the host between the running VMs.

3.6.5 Conditions, Objects and Order of Testing

The main objects of testing are the analyzer and the configurator applications that work with applicative models of information processes and implement the methods previously proposed for automatic configuration of the VM. Application prototypes in Python 3.5 have been developed for a simple functional test of the proposed algorithms. The test environment is a physical server with parameters presented in Table 3.3.

Two VMs run on the Azure server. Initially, VM resources are allocated as shown in Table 3.3. An Apache Web server with an MS SQL database runs on each of the VMs with a single Web page and a unique URL for queries. The first VM emulates

Table 3.3 The parameters of the testing environment

Parameter	Value	Comment
Cloud platform	Microsoft Azure Virtual Machine	Class General purpose, Standard_A4 [25]
Shared database	100 units eDPU, storage size 100 Gb per pool, 200 databases per pool	
Parametres of VM		
Operating system	MS Azure Linux Server	
RAM	8 Gb	
Local database	2 Gb	
Disk	64 Gb	
Additional disk storage	32 Gb	For instances of managed disks there is a surcharge on transactions
CPU core	4	

an online store, where user's requests for transmission of reference information and order registration are processed. The second virtual machine emulates the operation of the website of a 3PL operator, to which the specifications for the transportation and storage of cargo are passed. The issue of optimizing resources is particularly relevant to fields such as these, primarily due to the volume of data to be processed [39].

The standard linux system log (syslog from the file/var/log/messages) with the logging categories kern, user, daemon and with all types of priority, except debug (levels 0–6) was employed as a logger running on each VM. The resulting logs were enriched with information about the caller. Later this enhancement will be automated.

The resource scheduler running on the host machine used Azure Stretch DB to manage the resources of the shared database, Additional Storage Access to manage additional VM storage, and the Azure Load Balancer to monitor the network connections. The application performance and intensity were demonstrated for the VM basic configurations. They were compared with the performance of the same applications running on the same VMs and the resources scheduler additionally required a configurator based on the modeled business processes. A full list of features is shown in Table 3.4.

The analyzer runs on the host machine. The application is executed by the user after collecting and post-processing of the VM logs. The operation algorithm is based on the methods described in [40–42]. A detailed description of the model is presented in [43]. The application uploads the results to the host machine as an XML file that contains the model of the information process corresponding to the serialized input log.

Table 3.4 Assessed features

Feature	Unit	Effect
The amount of unused space in DB Stretch	eDPU	Medium. Affects the need to purchase additional space in the DB Stretch
Frequency of access to additional disk storage	Number of transactions	High. Affects the cost of maintenance
The recording time of the user's documents to disk	ms	High. Affects the waiting time of the user
The response time of the website for synchronous (interface) user requests	ms	Critical. The key function of the system

The model of the process must meet the following criteria:

1. Any reduction of the process term must reproduce all sequences of calls recorded in the system logs. A variable with an assigned type is used as input. It corresponds to the types of user queries. Internal query structure is not modeled
2. The model of the process should not generate reduction chains contrary to the available log; however, it can produce chains that have not been recorded on the virtual machine
3. The memory of the shared database and any additional storage required at each step of the process is computed by means of types assigned to the transmission and reception data channels

The model is constructed automatically; the user can modify it to exclude anomalies of execution, add or clarify the semantic markup, and eliminate semantically redundant blocks. The model processes allow for analysis of the steps and resources required. Based on the process model, the scheduler application builds a configuration. The scheduler treats any VM as if it is performing a single information process. The configuration produced by the scheduler applies only to the parameters, which can be managed without changing critical features of Azure VM: access to additional storage, shared database, and the load balancer configurations. This configuration is computed by the algorithm of sequential reduction of process terms.

The algorithm reduces the term for each of the parameters that trigger its initial step. For each reduction, the algorithm compares the resources with the values from the previous runs and selects the maximum. The list of such maximum values is the optimal configuration.

3.6.6 Analysis of Testing Results

The system was tested for load simulating the user activities [44]. The test data were based on real examples of queries to the system similar to the ones deployed on the same VM. The control parameters for testing were measured in three sessions of testing. The test results for VM parameter optimization are presented in Table 3.5.

As the table demonstrates, the application allowed determination of the configuration that minimizes the resource costs of the VM server. It is possible to optimize the parameters of accessing the shared database and additional disk storage.

The analyzer and the scheduler implement the optimization algorithm [36]. This algorithm is based on applicative modeling of information processes performed by the VM.

The scheduler implements the optimization algorithm presented earlier in this chapter. To simplify implementation and testing, the initial information process model is implemented by the analyzer. The applicative approach advantages are direct transfer of the formal model into code, immutability of objects, and versatile i.e. agile type system [45]. Using the formal applicative model allowed accurate specification of the application testing and verification of the outputs for the known input data. Test cases represent the π-calculus terms and the algorithm loaded into the abstract machine [46]. The output of the abstract machine was compared with the configurator.

The input data for the configurator was based on the model of information process. The terms of the VM processes are presented in Fig. 3.23.

In these terms, types of transmission and reception data channels correspond to the resources. The same data types are available for other resources. The first process describes two query types: for a Web form and via FTP. The second process describes receipt of a query via a Web form and its further processing by an online store. When modeling the information processes this way, we are unable to describe an asynchronous process communication; however, such an extension can be added in the future [47].

Table 3.5 The results of experimental verification of the system

Parameter	Change of value (%)	Comment
The volume of unused space in DB Stretch	−5	
Frequency of access to additional disk storage	−5	
The recording time of the user's documents to disk	−0	No significant difference in the recording time to disk
The response time of the website for synchronous (interface) user requests	−0	The algorithm of the load balancer Azure VM, and the test case did not give the ground to track the impact of the system over network resources

The process of a virtual machine (3PL)

$$a(x).b(z).\nu l_1.\nu l((l\bar{x} \mid l(x_1).F(x_1)[l_1,z] \mid l_1(z_1).F_1(z_1)[i,k] \mid F_1(z)[i,k])) \mid i(k).\mathbf{0}$$

$$\Delta(a{:}\,\alpha[1-20], b{:}\,\beta[200-800], l_1{:}\,\beta, l{:}\,\alpha)$$

The process of a virtual machine (online store)

$$a(x).\nu l_1.\nu l.\nu l_2((F(x)[l,y] \mid l(y).F_1(y)[l_1,z] \mid l_1(z).F_2(z)[l_2,q] \mid l_2(q).b\bar{q})) \mid i(k).\mathbf{0}$$

$$\Delta(a{:}\,\alpha[1-20], l{:}\,\tau[2-5], l_1{:}\,\tau, l_2{:}\,\tau_2[1-2])$$

a – channel to retrieve data via HTTP(S); ***b*** – channel for receiving data via FTP; $\boldsymbol{l, l_1, l_2}$ – channels of communication within the servers (RPC calls, shared folders on the server, etc.), ***i*** – technical channel to signify the completion of the process, $F(x)[l,z]$ – the function of the data unit ***x***, data transmission in the name *z* through the channel ***l***.
$\boldsymbol{x, y, z, k}$ -- associated data units.

Fig. 3.23 The terms of information processes of VM

The process starts by getting the names of the operations that receive queries from the client. The applicative approach allows to avoid detailing of the internal structure of each individual query. Instead, the query is matched with the name of a certain type, which is essential for the experimental verification [48].

According to the testing method, we used the model of input representing the process

⟦!i^(n_1)⟧_1 (x_1).⟦!i^(n_2)⟧_2 (x_2)... ⟦!i^(n_m)⟧_m (x_m),

where each elementary process

⟦!i^(n_m)⟧_m (x_m)

corresponds to the type of data items transmitted to the server and recorded in the log. The type assigned to these names describes the interval of the query size in Mb and uptime of network connections required to transmit the request to VM. The evaluation context of the abstract machine contains variables for resource optimization. The abstract machine performs step-by-step reduction and changes the resource context.

Our realistic applicative model allowed scheduling the hypervisor of a host machine based on its actual operations rather than abstract time moments. This schedule is better for geographically distributed systems and cloud platforms. It sets a system administrator free from adjusting timezone changes.

In this section, we have considered the application of previously developed algorithms of VM automatic configuration deployed on the Microsoft Azure cloud server. To verify the approach, we created and tested the model.

The tests showed that methods of the automatic configuration allow for optimizing settings of multiple VMs and the hypervisor.

The development of the proposed methods and applications would be the transition to configuring a local server instead of cloud VM for access to a greater number of parameters, which can be distributed between running VM (especially RAM, swap, CPU time). This will allow a full-scale comparison between the solution with an inaccessible physical layer for a greater overlap of shared resources [49], and a local but more detailed physical solution, as well as adapting the existing techniques of resource allocation by the hypervisor between VMs to the applicative approach.

Further development of the algorithm would include: (i) improved communication of VMs/server processes and data, (ii) enriched VM logging, and (iii) optimizing multiple independent processes.

This will allow for more accurate modeling for such instances as inaccessible physical layers. Further algorithm development would embrace more intensive data/process communication, more detailed system logs, and multiple business processes. This would make the solution even more agile.

3.7 Conclusion: How Services Work

In this chapter, we have addressed customer-related issues of mission-critical software development, which added agility to the general framework of the lifecycle concepts and languages discussed earlier. To efficiently focus on the client's side, we included interfaces and services into the crisis manager's toolkit. Therewith, we cross-examined a few types of the agile service-based approaches such as SOA, grids, clouds and Microservices. For each of these types, we did a SWOT analysis from the standpoint of agility.

In order to add value to the agile software development concepts of Chap. 1, we focused on the principles and patterns of service-based approaches and their application to software development (Chap. 4 will give more details on patterns and practices). We recommended service adjustment for crisis and mission-critical software development. The evolution of the service-based approaches—from SOA to grids, clouds and Microservices—resulted in better agility in terms of efficient management of big and heterogeneous data. We reinforced our analysis by a set of case studies on the service-based approaches that included banking and financials, customer relations, and geodata management. We supported these case studies by the state-of-the-art architectures such as virtual machines, clouds and Microservices.

Besides the family of principles, patterns and practices, the final cut of the crisis-responsive development of mission-critical software should include a set of human-related factors that might dramatically influence agility of the knowledge transfer and target software product. Our previous case studies only gave quick hints on these human factors; the final chapter of this book and [4] contain more details.

References

1. Foster, I., et al. (2008). *Cloud computing and grid computing 360-Degree compared.* Retrieved November 08, 2017, from https://arxiv.org/ftp/arxiv/papers/0901/0901.0131.pdf.
2. Larry Ellison's Game Plan. (2009). Retrieved November 08, 2017, from https://www.forbes.com/2009/09/22/oracle-zander-sun-intelligent-technology-ellison.html#5e2687b347d8.
3. West, D. (2016). *Essential micro-service architecture*. Retrieved January 18, 2018, from http://2016.secr.ru/lang/en/master-classes/essential-micro-service-architecture.
4. Zykov, S. V. (2016). *Crisis management for software development and knowledge transfer*, Issue 61, 133 p. Springer Series in Smart Innovation, Systems and Technologies. Switzerland: Springer International Publishing.
5. MSDN Microsoft. "*Chapter 1: Service Oriented Architecture (SOA)," MSDN*. https://msdn.microsoft.com/en-us/library/bb833022.aspx.
6. Rossi F. D., de Oliveira I. C., De Rose C. A., Calheiros R. N., & Buyya R. (2015). Non-invasive estimation of cloud applications performance via hypervisor's operating systems counters. In *The 14th international conference on networks (ICN), 2015, IARIA* (pp. 177–184).
7. Rivera, J. (2015). *Gartner identifies the top 10 strategic technology trends for 2015*", *Gartner*. http://www.gartner.com/newsroom/id/2867917.
8. Fowler, M. *Microservices a definition of this new architectural term*. http://martinfowler.com/articles/Microservices.html.
9. JON BRISBIN. (2013). *Reactor-a foundation for asynchronous applications on the JVM*. (Spring) Retrieved from http://spring.io/blog/2013/05/13/reactor-a-foundation-for-asynchronous-applications-on-the-jvm.
10. Richardson, C. (2015). *Building microservices: Using an API gateway*. Retrieved from https://www.nginx.com/blog/building-Microservices-using-an-api-gateway/.
11. Nginx. *NGINX LOAD BALANCING–HTTP LOAD BALANCER*. Retrieved from https://www.nginx.com/resources/admin-guide/load-balancer/.
12. Craig Williams. (2015). *Is REST best in a microservices architecture? (Capgeminin)*. Retrieved from http://capgemini.github.io/architecture/is-rest-best-Microservices/.
13. Richardson, C. *Pattern: Database per service*. Retrieved from http://Microservices.io/patterns/data/database-per-service.html.
14. Richardson, C. *Pattern: Event-driven architecture*. Retrieved from http://Microservices.io/patterns/data/event-driven-architecture.html.
15. Richardson, C. *Introduction to event sourcing and Command Query Responsibility Separation (CQRS)*. Retrieved from https://github.com/cer/event-sourcing-examples/wiki/WhyEventSourcing.
16. Commonwealth of Kentucky. (2006). *NASCIO recognition award application: Enterprise service bus*.
17. The ESB in the Land of SOA. (2005). *Information technology research: Enterprise strategies.* Boston: Aberdeen Group, 7 December 2005.
18. Public Cadastral Maps. https://pkk5.rosreestr.ru.
19. Open Data Portal. http://data.gov.ru/frontpage?language=en.
20. Xiong P., Wang Z., Jung G., & Pu C. (2010). Study on performance management and application behavior in virtualized environment. In *Proceedings of the In Network Operations and Management Symposium (NOMS), 2010* (pp. 841–844). IEEE.
21. Ocean M. J., Bestavros A., & Kfoury A. J. (2006). snBench: Programming and virtualization framework for distributed multitasking sensor networks. In *Proceedings of the 2nd International Conference on Virtual Execution Environments* (p. 89). ACM.
22. Chowdhury, N. M., & Boutaba, R. (2009). Network virtualization: State of the art and research challenges. *Communications Magazine, 47*(7), 20–26.
23. Schaffrath G., Werle C., Papadimitriou P., Feldmann A., Bless R., Greenhalgh A., & Mathy L. (2009). Network virtualization architecture: proposal and initial prototype. In *Proceedings of the 1st ACM Workshop on Virtualized Infrastructure Systems and Architectures* (pp. 63–72). ACM.

24. Raghavendra, R., Ranganathan, P., Talwar, V., Wang, Z., & Zhu, X. (2008). No power struggles: Coordinated multi-level power management for the data center. *ACM SIGARCH Computer Architecture News, 36*(1), 48–59.
25. Milner, R. (1982). *A calculus of communicating systems*. New York: Springer.
26. Boudol, G. (1998). The π calculus in direct style. *Higher-Order and Symbolic Computation, 11*(2), 177–208.
27. Milner, R., Parrow, J., & Walker, D. (1992). A calculus of mobile processes. *Information and Computation, 100*(1), 1–40.
28. Pierce B., & Sangiorgi D. (1993). Typing and subtyping for mobile processes. In *Proceedings of Eighth Annual IEEE Symposium Logic in Computer Science. LICS'93* (pp. 376–385).
29. Dillon T., Wu C., & Chang E. (2010). Cloud computing: Issues and challenges. In *Proceedings of the 24th IEEE International Conference Advanced Information Networking and Applications (AINA), 2010* (pp. 27–33).
30. Chowdhury, N. M., & Boutaba, R. (2009). Network virtualization: state of the art and research challenges. *Communications Magazine, 47*(7), 20–26.
31. Gromoff A., Kazantsev N., Shapkin P., & Shumsky L. (2014). Automatic business process model assembly on the basis of subject-oriented semantic process mark-up. In *Proceedings of the 11th International Conference on e-Business ICETE, 2014* (pp. 158–164).
32. Melekhova A., & Vinnikov V. (2015). Cloud and Grid, Part I: Difference and Convergence. *Indian Journal of Science and Technology*, *8*(29).
33. Melekhova A., & Vinnikov V. (2015). Cloud and Grid, Part II: Virtualized Resource Balancing. Indian Journal of Science and Technology, *8*(29).
34. Poon, W.-C., & Mok, A. K. (2010). *Bounding the running time of interrupt and exception forwarding in recursive virtualization for the x86 architecture* (Technical Report VMware-TR-2010-003). VMware, Inc., 3401 Hillview Avenue, Palo Alto, CA 94303, USA.
35. Melekhova, A., & Markeeva, L. (2015). Estimating working set size by guest OS performance counters means. *CLOUD Computing*, *48*.
36. Waldspurger, C.A. (2002). Memory resource management in VMware ESX server. *ACM SIGOPS Operating Systems Review*, *36*(SI.), 181–194.
37. Zykov, S., & Shumsky, L. (2016). Application of information processes applicative modelling to virtual machines auto configuration. *Procedia Computer Science, 96,* 1041–1048.
38. *Economics Paradigm for Service Oriented Peer-to-Peer and Grid Computing*. Retrieved November 7, 2017, from http://www.cloudbus.org/ecogrid/colostate.html.
39. Milner, R., Parrow, J., & Walker, D. (1992). A calculus of mobile processes, i. *Information and Computation, 100*(1), 1–40.
40. Zykov, S., & Kukushkin, A. (2013). Using web portals to model and manage enterprise projects. *International Journal of Web Portals, 5*(4), 1–19.
41. Van der Aalst, W. M. (2011). *Discovery, conformance and enhancement of business processes*. Springer.
42. Shumsky, L., et al. (2013). A synthetic approach to building a standard model of subject areas in the integration bus. In *3rd international symposium on ISKO-Maghreb* (pp. 1–7).
43. Gromoff, A., et al. (2014). *Automatic business process model assembly on the basis of subject-oriented semantic process mark-up* (pp. 158–164). SCITEPRESS-Science and Technology Publications.
44. Shumsky, L. (2014). Semantic tracing of information processes. *Software Systems And Computational Methods, 1*(1), 80–92.
45. Draheim, D., et al. (2006). *Realistic load testing of Web applications* (pp. 11–70). IEEE.
46. Barendregt, H. (1991). Introduction to generalized type systems. *Journal of Functional Programming, 1*(2), 125–154.
47. Berry, G., & Boudol, G. (1992). The chemical abstract machine. *Theoretical Computer Science, 96*(1), 217–248.
48. Boudol, G. (1992). *Asynchrony and the pi-calculus. Rapports de recherche- INRIA*. Institut national de recherche en informatique et en automatique.
49. Barendregt, H. P., Dekkers, W., Statman, R. (1977). *Typed λ calculus* (p. 618).

Chapter 4
Agile Patterns and Practices

Abstract This chapter focuses on how agile best practices depend on human-related factors; we integrate this approach with design patterns and see how that can be used to improve the overall agility. We use the same high level layer-based pattern as we did for languages in Chap. 2 and for services in Chap. 3. We discuss the concept of knowledge transfer and provide a case study on the key factors which promote this transfer in crisis. We look at how these factors can be applied to a Russian startup—Innopolis ecosystem, and how the project can be improved with more flexibility. Further information on knowledge transfer is given through a case study on open education resources including massive open online courses. Finally, we discuss patterns and anti-patterns in crisis to improve the agility and efficiency of software development in mission-critical systems. Our discussion embraces agile transformation of the software system for air communication management, which is a human factor-dependent domain.

Keywords Best practice · Knowledge transfer · Layer-based approach
Design pattern

4.1 Introduction: Why Agile Patterns and Practices?

In the previous chapters, we have moved from the general concepts of agility and crisis to more elaborate aspects of software development such as languages and services. A software development methodology typically includes a set of formal models, processes, methods and tools. However, the final steps of software development and implementation including detailed artifacts and best practices of agile development were generally out of the scope of the discussion. This was so, because these agility practices essentially depend on human-related factors, which were reserved as the primary subject of this chapter.

The other vital aspect of the chapter is software design and development patterns. To build crisis resistant software, we suggest strategic reuse of high level patterns. The starting point is the design pattern approach introduced by Erich Gamma and further enhanced by the famous "Gang of Four" [1]. We will discuss the practices

S. V. Zykov, *Managing Software Crisis: A Smart Way to Enterprise Agility*,
Smart Innovation, Systems and Technologies 92,
https://doi.org/10.1007/978-3-319-77917-1_4

that make certain software development (and knowledge transfer) patterns efficient in crisis. As we already know from Chap. 3, which discussed the Microservices, the choice of specific patterns may help or hinder agile software development. This observation gives rise to the anti-patterns, or typical pitfalls to avoid in crisis.

Of course, the correct pattern choice is not the only delimiter for crisis resistant software production. The choice of patterns and anti-patterns critically depends on human factors and soft skills related to these factors such as: negotiation, communication, and teamwork. In crisis, the human factors are mission-critical as they may promote or inhibit knowledge transfer for the future software product between client and developer sides. These two parties typically represent organizations rather than individuals, i.e. they are complex business structures. Therefore, knowledge transfer is a complex multi-sided and multi-staged process. We cross-examine this knowledge transfer process in multiple contexts, which include: ecosystem development, software engineering curricula transfer, and knowledge management in the open educational resources (OER). The case studies are based on [2].

This chapter is organized as follows. Section 4.2 presents a case study on knowledge transfer and discusses how resilient communication improves its agility despite mentality and other diversity barriers. Section 4.3 gives a framework for OER knowledge transfer, which includes metadata (i.e. knowledge) management architecture. Section 4.4 analyzes pattern-based agile transformation of a large-scale system in such a human factor-dependent subject area as aircraft communication management. The conclusion summarizes the results of the chapter.

We focus on patterns and practices that improve software agility in crisis.

4.2 Agile Knowledge Transfer

To accommodate the Russian economy to the new digital era, the following issues should be addressed [3]:

(i) Improving technological infrastructure
(ii) Establishing adequate legislation for intellectual property
(iii) Enabling efficient collaboration between public and private sectors
(iv) Increasing the number of skilled IT professionals
(v) Improving IT education quality.

Innopolis started in 2012 as an ambitious project to solve the above problems. In 5 years, the situation changed dramatically from multi-channel financial support to predominantly self-sustaining orientation. This self-sustaining goal is probably the biggest challenge, as the expenses are growing rapidly with development and many of the outcomes are still in progress.

The city of the future is located nearly 40 km (25 miles) from Kazan, the Tatarstan capital, close to the famous Russian Volga River, neighboring all-year ski resort, the golf course, where the Russian President Cup is held yearly, and the Olympic shooting stadium.

The special economic zone, separate from the University campus, is around 200 hectares. Its potential residents are innovative companies, preferably doing IT. The residents will get a 5-year zero tax for land and profits, a 5-year zero tax for assets and transportation, and social tax of 14%. The conditions are attractive; however, they are not accessible to everyone.

The current number of Innopolis residents exceeds 30. These include a number of major Russian IT companies, such as New Cloud Technologies (the developer of My Office, a Russian software toolkit), the National Center for IT, a subsidiary of the Rostech state enterprise, the IT developer and communication operator for the Football FIFA 2017 Confederations Cup Russia and the 2018 FIFA World Cup.

In 2016, Sberbank Technologies, a subsidiary of Sberbank, started partnership with Innopolis. In 2017, this IT company opened a branch office here for 600 employees; some of those will be relocated from other offices. The same year the Yandex Company became an Innopolis partner; it will develop the famous search engine here. Another large partner is the Voskhod research institute, a developer of innovative governmental IT solutions; they are opening the Regional R&D center in Innopolis. Kaspersky Lab is one more example of a famous resident; the company opened a training center for cyber defense based in the Innopolis University (IU).

Since 2016, Innopolis is run by the local Tatar government. The Tatar Ministry of Communications will keep registry of residents, control the agreements and act as a major client.

The initial plan was to reach a 150,000 inhabitants milestone in 20 years. The linear projection gives the desired current number as around 40,000 in 5 years. However, the current number of inhabitants is around 1,500–2,000, according to various estimates. The Official House of the Mayor of Innopolis is a 6-floor residential building; one level is for the administrative offices. The Innopolis City Mayor is 36 years of age and wears jeans; he was elected in 2016 (a possible reason for this was that the local Tatar government was dissatisfied by the innovative city growth).

At the end of 2016, the prior Innopolis City Mayor, Yegor Ivanov, elected in 2014, resigned. Roman Shaikhutdinov, the Tatarstan Minister for Information and Communication, said that the Mayor completed the tasks he had been assigned, and he is to return to Moscow. The experts, however, were unanimous that changing the Mayor will not accelerate city development as the results will not be noticeable before 10–15 years.

Innopolis city was designed by the RSP Architects Company from Singapore headed by Liu Tai Ker, who proved that any well-planned city is comfortable for living. Innopolis has every facility accessible on foot within 15 min, these include university and campus, sports gym, clinic, techno park, and the school. The residential sector consists of 16 buildings which can host around 4,000 people (the initial planned capacity of the city is 150,000 people). In 2016, an additional 105 apartments in two houses were built. Each of the apartments has 1 or 2 bedrooms and is fully equipped with furniture or unfurnished. All the apartments are for rent only (except for a very few special arrangements with the top level faculty members). To attract and keep inhabitants, monthly fees were reduced twice and currently are around $100–150. All houses have underground car parking (included in the above rent fee); however, remote access to the gates does not operate properly in this innovative city. Another option is a town house with a 12%+ yearly mortgage. The town house space starts from 114 sq.m, the price per sq.m is around $1,000 (a fully furnished facility would be 12% more).

Innopolis city was founded in 2012 as the Russian IT capital and officially opened to the public in 2015. Currently, the city is in search of new inhabitants. The local Innopolis population is only around 2,000 with average inhabitant's age of 27 (as the students make up nearly half of this); nearly the same number of people travel daily from Kazan, the Tatarstan capital.

Innopolis features a safe city environment with modern infrastructure, friendly ecology and a set of facilities for learning and professional development.

The city economics is high-tech based. The ecosystem includes learning facilities with the first Russian 100% IT University and business infrastructure for IT residents with attractive living conditions.

Innopolis is the youngest Russian city to become the IT industry's capital. The planned capacity is 80,000 inhabitants involved in IT, the university will scale up to 5,000 students from its current 550. The 40 Gbyte courseware repository probably contains little or no rich media, as its size allows for as little as 40 h of video lectures of MP4 quality in total.

As we mentioned, the IU is the brain of the future city, its annual budget is almost RUR 1bln, or EUR 150M. In 2016, the teaching professors and researchers published 113 papers in total (or 1.6 papers per person), which still looks rather low (whereas the previous year ratio was 1.3). In 2015, the IU income has almost doubled as compared to the previous year (RUR 378M vs. RUR 755M in 2014); however, the number of the students (and professors) also doubled.

In 2016, the IU yearly expenses per student exceeded RUR 1.5M (or $25,000), which is almost 10 times better than the local favorite, Kazan State Federal University with around $30,000/year. Of these, nearly 50% is the professors' salary, which is doubtful to keep up with at the current IU growth rate that doubles every year. Due to the crisis, the IU sponsors' donations dropped down from RUR536M (approx. $8.9M) in 2015 to RUR337M (approx. $5.6M) in 2016 (see Fig. 4.1).

The doubling of researchers did not income contributions to any funding in 2015, although their number exceeded the number of teaching staff by 150%. This resulted in University revenues being reduced by 90% for this period. Consequently, the research staff were sharply reduced in 2016 so that it became equal in number to the teaching staff. Perhaps, the issue was increasing the research staff number paying less attention to quality. Of the current 48 IU researchers, the top three boast an H-index of 20+, while the top nine have a moderate 10+ and the rest are not reported. Thus, average IU researcher's H-index is probably below 5 points. The faculty members generally do not contribute research-based funding (the top seven have a moderate H-index of 5+); however, their salaries are some 10% larger. Since IU failed recruiting top level researchers, the grant projects number and revenues decreased in 2016 approximately 1.5 times and two-fold respectively.

The number of laboratories increased following the new trends (such as big data, cybersecurity, digital technologies etc.). However, none of the inefficient existing

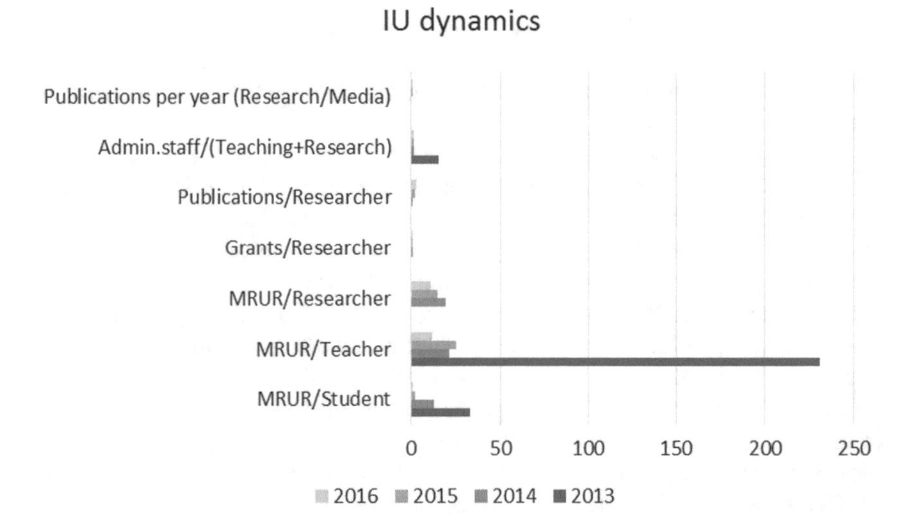

Fig. 4.1 IU dynamics (2013–2016; source: IU 2017 annual report)

laboratories were closed. Perhaps, it was this slow reaction to research inefficiency that resulted in a sharp dropdown of the grant projects, both in number and value. Meanwhile, the bureaucratic part of the university is relatively large and continuously growing. The administrative staff amount to nearly 60% of the total staff number. In 2016, a small reduction of this overhead made the administrative staff 57%, which seems still too high for the current crisis situation.

The PR department reports a large amount of publications in the media news; these dramatically outperform the research publications. The latter make a minor share of 4–5%. Essentially, these media news require a certain budget; however, they do not seem to bring money to the university.

In the current crisis situation of decreasing budget and stiff competition for the grant projects, the University has still made impressive progress. In 2016, the Innopolis University launched an academic exchange program for students. The number of student exchanges is relatively low; however, the level of the hosting universities is quite competitive. One example is South Korea Advanced Institute of Science and Technology (KAIST), which is in the top 50 computer science universities. Concerning internships, several IU students joined a world renowned European Organization for Nuclear Research, the famous CERN. This would not be a surprise for Russian MEPhI, the National Research Nuclear University, which has a fantastic team of very strong nuclear physicists and enjoys an extensive and long term cooperation with CERN. For the Innopolis University, whose focus is 100% IT, this new CERN partnership is an impressive sign of good progress.

Additionally, new partnerships were established with nine universities, including those in EU and China. Most of these are second level universities; however, the IU is already a partner of a number of top level technical schools. Moreover, in crisis any prospective partnership is mission-critical. Another important sign of progress is a European Union Erasmus + grant for academic mobility with the Dublin Institute of Technology in Ireland. In this project, the partners plan to launch a joint research lab.

Currently, the IU focus shifts from narrower domain of Software Engineering to a much wider area of Computer Science. Therefore, the number of disciplines taught is rocketing; during the 2016/17 academic year it increased 2.5-fold and now includes as many as 84 courses for bachelors, of which 56% are core, and the rest 44% are the electives. Of these electives, nearly one third is focused on "soft" skills, which include entrepreneurship and communication; this is mission-critical in a crisis. The year 2016 was also a start for individually selected educational program trajectories. The choice for the final year bachelors includes 27 electives, of which each student should choose four of his or her own preferences. For master students, the approach is similar, it forms an innovative and flexible academic environment and helps to cope with the crisis. In total, the Innopolis University boasts an EU-compatible ECTS catalogue which includes 161 courses. For each course, this directory lists the number of ECTS credits, majors, target audience, academic year, semester, and key words, to name a few. This catalogue helps the students to proactively monitor the curricula and to manage it efficiently in terms of electives and individual learning curves. Not only did the programs and courses increase in number, they also became much more business focused. Therefore, expert opinions were acquired from the leading digital industry companies and the major HR agencies in order to understand the fields and occupations of current and prospective demand. IU built its own course directory and developed a centralized courseware repository.

Concerning company-to-university cooperation, the major IU partners are Gazprom, the largest Russian oil-and-gas enterprise, Aeroflot, the top Russian airline, and Yandex, the most important Russian internet company.

To provide a better course climate, the Innopolis University has set up a policy containing a number of ingredients. These include: a cognitive knowledge transfer loop, a centralized courseware repository and directory, and feedback-based techniques to enhance communication [4, 5]. In crisis this course climate policy becomes a matter of primary importance. That is why we describe each of these ingredients in details below.

The cognitive loop includes the following phases: "Plan-Do-Monitor/Assess-Adjust". We describe this in more detail in the previous book on crisis [2]. The recent innovations that added to this general framework were inherited by Innopolis from the Master Program in Software Engineering of the Carnegie Mellon University. These innovations were framed together into the Education Quality Control System (EQCS). The EQCS applies to every teaching position of the University. It includes a dozen of correlated elements, which generally refer to assessment and feedback activities (see Fig. 4.2). To apply for any teaching position at the University, in addition to CV and other standard documents (recommendation letters, statement of work, course/project/publication list etc.) an open lecture is required for any candidate. After enrollment, successful applicants are assessed in different ways. The primary assessment is the on-site evaluation; this means spontaneous (i.e. unscheduled) attendance of certain lectures by the faculty and/or administration board members. Another way of assessment is an anonymous student poll on course quality; this happens at the end of any semester prior to exams. In addition, administration arranges mid-term meetings with the students, where the course instructors interact with the dedicated student representatives to receive their feedback and adjust/improve the course delivery accordingly.

In parallel to the mid-term meetings, the mid-term assessment of students' knowledge happens. This includes student performance evaluation, which assists in identifying the students who require special treatment. The forms of such treatment may be: personal mentoring, professional orientation, and social/psychological support, to name a few. The feedback from these mid-term activities is collected, processed and the improvement proposals/recommendations based on the above are delivered to the instructors and administration.

Other ways of feedback include: faculty reflection, email student support, testing cognitive skills, and monitoring student performance. Faculty reflection refers to instructors self-interviewing and self-adjustment; it is part and parcel of the knowledge transfer lifecycle. This happens continuously as a part of the closed-loop process [4, 5]; however, EQCS also includes it as a formal questionnaire, which typically happens after the examinations. A dedicated email-based service is an instant feedback source for the students concerning the courses they learn, current problems and challenges, and possible solutions and improvement proposals. This service has a single entry point known to every IU member: *education@innopolis.ru*, which is important for urgent matters i.e. in the events of local/individual crises. Since any questions regarding the educational process are welcome, no problems are hidden. In addition to the above-mentioned practices, this is also an agile crisis management tool. In progress of the semester, the students take tests for their cognitive skills. This is critically important to make an efficient and informed knowledge transfer between the instructors (as the transmitting side) and the students (as the receiving side) [2, 4, 5]. To keep the testing results available and consistent, the university uses a uniform platform of the Moodle learning management system (LMS). The author advised for this LMS and participated in its initial testing in 2014/2015 academic year for his course on Personal Software Process [2]. The LMS allows monitoring and tracking student activity and performance both in terms of the course content and the cognitive "soft" skills (see Fig. 4.2).

IU is the most populated and active place of the new city. The modern building with spaceship-like interiors is linked to student dorms so that they can reach it without going outdoors. This is essential as the winter here is usually harsh, cold and windy. The official teaching language is English—every student is at least intermediate, every faculty member must be proficient in it. Other student admittance requirements include: personal record of participation in competitions, contests and conferences; high grades in mathematics and IT, and basic programming skills. In 2016, the competition rate reached 300 people per one student position. The current student GPE is around 8.5.

Concerning geography, the University student cohorts have recently become international and therefore the community is culturally diverse. As of 2016, 313 citizens

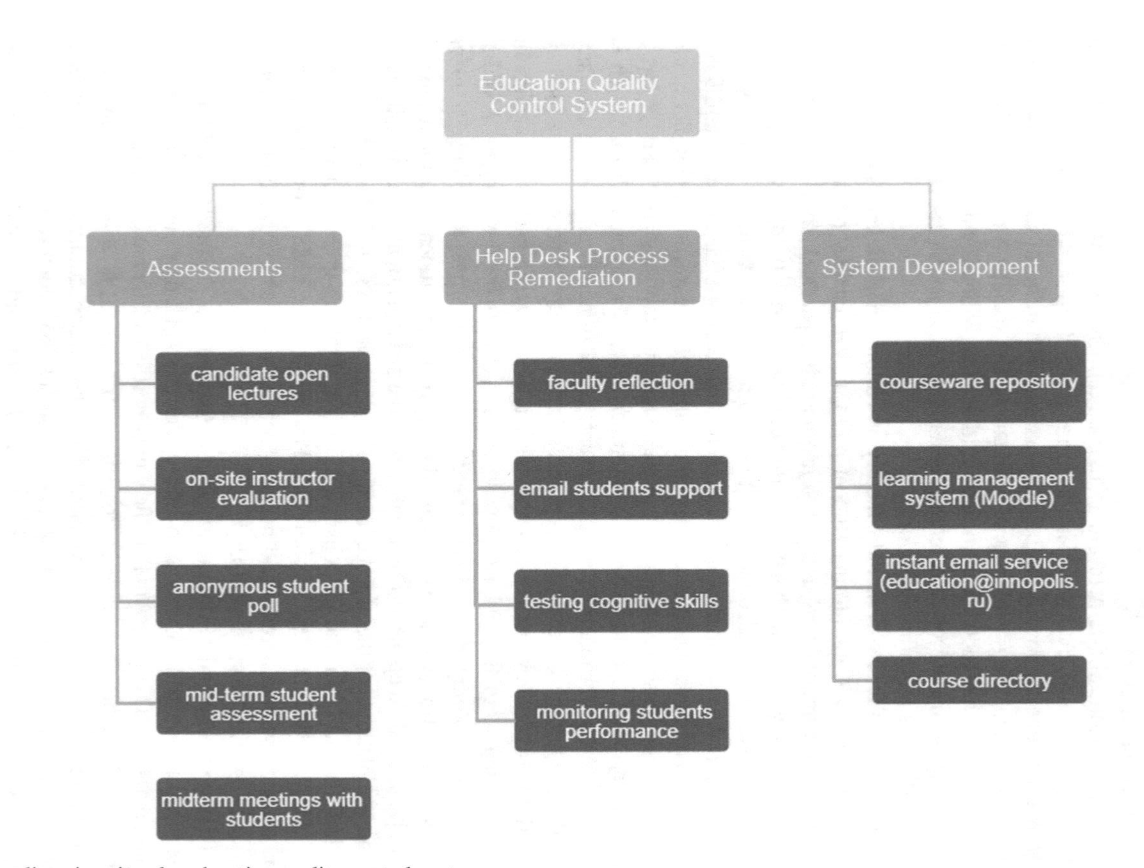

Fig. 4.2 Innopolis university: the education quality control system

of 10 countries became the IU students; the number of applications was 9,200. The faculty number grew to 45 (in 2015 there were only 23). The R&D staff is 48 people (in 2015 there were only 39). The administrative staff is 124 people (in 2015 there were only 93). IU includes 16 labs and R&D centers; the top IT experts and professors from Canada, China, Italy, and other countries teach there. The number of research publications in 2016 was 113.

After admittance the students receive a grant, which fully covers the tuition fee. Over 40% of the curriculum is project-based teamwork. For bachelors, the first two years the students get the basics in computer science and engineering. From the third year, they start specializations in AI, cyber security, software engineering, bioinformatics, game and application development, and big data.

The students enjoy timely, detailed and goal-directed feedback, and they feel constant personal improvement. In a crisis, this feedback is mission-critical. The students take internship with IU partner companies to join them upon graduation. Many students join IU with business startup ideas; the sponsors are often interested in the R&D based on these ideas. The IU sponsoring partners include the MTS and Megaphone mobile operators, and the Acronis and Parallels IT companies.

After the university classes, the residents (university students, faculty, and staff) can learn dancing, photography, Chinese or other foreign language (to name a few options) for free. The students enjoy free tuition and free access to any of the city facilities; moreover, they get a $300–600 scholarship depending on their grades; this amount essentially exceeds an average Russian university scholarship. However, the contract prescribes the graduates to work at a local resident company for at least one year; otherwise they have to pay the scholarship back to the university, which is nearly 50% funded by these companies.

In 2016, the Mayor said that developing entertainment infrastructure is a long-term matter. The EUR300M Futuroscope entertainment park with a French investor is to host up to 5,500 guests. Another long-term project is the EUR43M ropeway from Kazan to Innopolis. If these two projects merge, they have a good chance to produce a synergy. In 2016, as an addition to the Cava coffee house the first beer restaurant opened in Innopolis. The other assets of the kind include a children's entertainment center, gym, a beauty salon, and a hostel. The Tatarstan local government has a dedicated program on cultural and entertainment development for Innopolis; however, the current focus is Kazan cultural assets easily accessible by car.

There is a state-of-the-art medical center; however, its current load is rather low. The local inhabitants are served here for free; some of the patients come from remote Tatar districts. The experts of this clinic are available once or twice a week; one can apply by phone or internet. The most common way of communication for Innopolis people is the Telegram messenger. Local problems, bus schedule and other topics are discussed there; one can even chat privately with the IT Minister or IU rector.

One more city innovation is the "concierge service"—a single access point system for taxi reservation, bus schedule, apartment renovation and similar services. The living problems are quick and easy to solve here. The kindergarten is just across the road; the pre-school is combined with the elementary school, where the first year children learn English, IT and robotics.

The university uses dynamic feedback to provide "resonant oscillation" in terms of communication between students, instructors, mentors and stakeholders. Ingredients to support this resonant communication are: InnoBootCamp, Student Association, clubs, events, and competitions, to name a few. The Student Association has a Chairman, a Board of Founders, and members in charge of fundraising, PR and dedicated committees. The InnoBootCamp is a project that adapts new students to the educational process and other university activities. The idea is inherited from CMU Master Program in SE [2]. It is enhanced so that the young people build a strategy that helps get them involved in the university activities in the best possible way.

The other feedback enhancement ingredients include: IT competitions, reward system, media relations, and a set of IT events such as conferences, workshops, summer schools. Of the 35 recent country-wide competitions, the IU students were prize winners in nearly 60%. These included: AI FinTech Hackathon, IX Open Olympics in Programming, Microsoft Hackschool, FashionHackathon, InspiRussia Hackathon, HackDay, BattleBots by Yandex.Money, and Hackathon on Intelligent Transport Systems and Human Mobility. The competition names show that the primary focus of the University is application development rather than computer science. The major application areas of the students' expertise are: financial, transportation, robotics, and cybersecurity. Yet another source of feedback is media relations. The PR activities help sharing the common corporate values between the students and teaching/research staff, integrating new members into the team, getting to know more about the areas of growth and development. In 2016, the PR department reported releasing 2,177 media publications on the University, which is over twice that of 2015.

A part of the environment is hosting conferences, summer schools, competitions in robotics, (external) trainings, guest lectures (160 in 2016) etc. The students get involved in the events, which more than doubled as compared to the 2015.

The outcomes of the IU activities are such industry level implementations as Aeroflot business process re-engineering, new medical developments and a few others. Innopolis researchers developed several innovation projects that may dramatically change our life in a few years.

One of these projects is the first Russian mobile operating system (OS) named Sailfish Mobile OS Rus, which is to compete with Google and Apple products. In Russia, there are around 75M smartphones. Of these, 74% use Google and 20% use Apple operating system, the iOS. Therewith, such devices may spy users: they gather personal information (contacts, calls, photos, videos etc.) and behavioral data (such as applications installed, frequency of their use, internet sites visited, and the sequence of these visits). This is a potential threat not only for the individual but also for the state. This is the reason for development of the Russian mobile OS. The OS is based on the open source Sailfish platform by the former Nokia developer; however, this is 100% domestic product. The entire OS infrastructure is located in Russia. The data processed by the smartphones never go outside the OS perimeter. The OS source code is also stored in Russia. According to the developers, the Sailfish Mobile OS Rus devices will secure governmental, enterprise and individual data. In

2016, the company sold the first batch of the preordered 1,000 Russian OS-based smartphones. The sales plan for 2017 is 500,000 devices under the INOI brand.

Another Innopolis resident, InspiRussia, develops blockchain technologies that become widely known due to Bitcoin. Blockchain is a secure data storage technology, which guarantees that an untrusted person will not be able to access the data. Thus, blockchain solves the problem of trust by eliminating the third party (e.g. a bank broker, depositary personnel etc.) from the process chain (e.g. buying stocks). InspiRussia develops blockchain solutions for Tatarstan governmental organizations. There are several possible applications of the technology. One example is land registration, where blockchain will decrease violations and improve property accounting. Some people avoid taxpaying as the accounting is poor. If the property database uses blockchain, the taxpayers will get a better service, and the budget will get more income.

Another example is a document management system, where blockchain eliminates a number of third parties which identify an individual or organization. This saves time and financial resources. Finally, blockchain helps in voting; it makes the system totally transparent and non-intrusive, as there is no single denial point. During elections, the blockchain will solve violation problems and assist in monitoring.

The Sputnik company presented one more Innopolis project, which will secure apartment blocks, assist in housekeeping and improve fire services. This is an integrated communication and surveillance device, which records everything in the proximity of the house. For instance, while the fire brigade is on the move to terminate a fire, the device warns the residents, and they are out the house before the brigade arrives. This saves valuable time. The device can be integrated with nearly any other hardware by means of internet of things (IoT) technology. For instance, it can be linked to a water leakage sensor; the householder monitors it and is sure that the situation is under control. In 2017, the device was shown in Moscow at the MIPS international exhibition on hardware for safety, security and fire prevention.

The Roadar Company is another Innopolis resident that develops computer vision. The startup outcomes will be used for car number recognition, and in such industries as healthcare, banking and mass media. The computer system for car number plate recognition is already available as a product. The idea itself is not novel; however, the recognition quality is much higher than that of the market competitors, the company reports. The system accurately recognizes dirty and hard to read number plates. This system can be used at car parking lots, exits, road surveillance cameras to monitor speed, violations etc. Currently, Roadar is working on a monitoring system for pedestrian crossings. In addition to these application areas the system may substitute a human for product quality control.

Yet another product will help students to find a job in their area of competence, and tell universities how to improve their curricula to match the market demand. This is the Career Portal product by the Innopolis subsidiary of the ICL company, a large-scale Tatarstan software developer. The portal is a meeting point for employers, students and universities. The portal stores student portfolios with a detailed competence list, internships etc. i.e. a database of approved knowledge and skills. This database will help to monitor the graduate's career (the individual demand will be rated), and universities will get independent ratings of learning quality from a number of employers and adjust accordingly. Career Portal is a rare project to be tested on the entire Tatarstan scale. The short-term plans include IT cluster-based testing. Later, the product will be adapted for the other Tatarstan industry clusters such as automobile, furniture, and food production, oil refining, and construction.

According to the 2016 annual report, the total UI expenses were around RUR1.0Bln ($17M), and each student learning cost was around RUR1.5M ($25,000). In 2016, Innopolis started planning for a new techno park which will cost around RUR2.1Bln ($35M). This will be a 5-level office building of 28,000 sq.m and over 900 workplaces. The project duration is one year. In 2016, a new water supply and purification system was installed.

Innopolis lacks clubs and cafés to relax after a busy day. The Tatarstan government and the city Mayor are working on that. Currently, besides the above mentioned cultural assets, there is a number of such places available in Kazan located some 40 km (25 miles) away.

At present, the University has reached its admission limit as the faculty staff is not ready yet (the planned capacity is 5,000 students). According to the IU top management, the main problem of the University is to attract the human resources. The city and the University have a giant potential; however, the Tatarstan alone cannot fill Innopolis with the employees. This should be done on a national scale.

4.3 Open Education Metadata Warehouse

The 'open education initiative' was founded on the fact that knowledge and education are public goods and that access to quality education is an essential human right [6].

The MIT's Open Courseware [7] stimulated many other institutions the world over to make available their educational resources. The idea surrounding the conception of digital online learning resources for (re)use was introduced into the public space since early 1990s. These *Open Educational Resources* (OER) comprise the learning content, tools and licenses [8, 9].

OER define educational materials as sets of resources, which have been collected, organized and described with the idea that they would be (re)used. Thus, the following appears to be the core of an OER: its purpose is to be reused or remixed, without cost. "Open" means that, for the end user, there is the elimination of technical obstructions (closed source code), little or no cost implication (subscriptions, licensing fees) and limited legal (i.e. copyright and related) issues [10].

As part of efforts to further extend the frontier of OER, there came the notion of the MOOC. *MOOC* aims at providing free, better access to educational resources with a view to bringing down the costs associated with higher education. Presently, MOOCs have enjoyed huge attention from the media, entrepreneurs, educationists and technologically savvy sections of general public [11]. This has spurned on elite educational organizations to put their course resources online by implementing e-learning platforms, such as edX. In partnership with prestigious institutions, new startups like Udacity, Coursera, etc. have been launched [7]. Expectedly, with a transition from open access to open educational resources, and of recent, open online courses, there is a new vigor and momentum among educational organizations to participate in this "open" movement. Therefore, this section explores concepts such as metadata, open education, data warehouses, with an emphasis on how they can together provide agility for high-quality educational resources.

OER, as an unfolding concept, faces certain challenges including availability, access, costs and quality of information on the internet. And, owing to the increasing growth of internet technology and the opportunity it offers to improve access and transfer knowledge to a wide range of users, there is a need to address some important issues that relate to better accessibility and discoverability by concentrating on the metadata elements and components of Web-based educational resources.

Significant attempts at broadening the reach and scope of education have been initiated by many open education promoters. For instance, in 2010, Stanford University came out with a MOOCs platform [11], and this move attracted scores of users, in their thousands, to free online courses. Since then, the MOOC concept has been enjoying a good measure of popularity and now includes offerings from many service providers in education via different MOOC platforms.

Despite the noble purposes behind this idea, it was found that OER has not been a smooth sail. This is because the community of users is faced with various challenges in harnessing in the potential. Today, huge collections of under-utilized learning materials are still residing on the internet. But the burden of locating those resources, evaluating their quality, connecting them to other materials, and sharing them with other prospective users usually falls on individuals.

This section aims to design a data-driven architecture that will facilitate agile access to quality educational resources.

Therewith, we are to develop: (i) an agile method to extract metadata from major e-learning collections; (ii) an efficient technique to classify extracted metadata into groups; (iii) a repository to warehouse the metadata; and (iv) a prototype of a OER metadata portal. As such, we adopt two major technical methods for managing metadata. The first one deals with text extraction from technical information. Second, the extracted metadata will be segregated into appropriate categories according to different criteria.

A rule-based approach will be applied in implementing the first of these tasks. Here, the approach will be of help in programming instructions specifying how to extract data from targeted platforms. As for the second aspect, template-based method will be introduced to classify metadata elements into classes based on similar characteristics. The contribution promotes agility by: (i) improving access to OERs by leveraging the power of MOOCs and other educational platforms; (ii) increasing awareness about the visibility of OERs on the internet; (iii) providing a medium for long term OER metadata storage.

Data Warehousing and Metadata Management

Data warehousing is a computing process used for getting data from source systems (legacy or operational), transforming them into organized and user-friendly formats and loading the resultant data into a designated data repository to encourage data analysis and ultimately support business decision-making [12]. During this process, the end user can, among other things, carry out ad hoc querying, generate reporting, and visualize data based upon certain criteria. The core objective of a data warehouse system is to produce a data repository and make data accessible in formats that are readily acceptable for decision support and other purposes [13]. A data warehouse system is made up of key mechanisms that need to be integrated and function together for efficient and effective operation. According to Tole [14], Kimball and Inmon describe the components/services that constitute a data warehouse system in relation to how they fit together, scale and grow. For warehouse design, Inmon suggests top-down approach whereas Kimball proposes bottom-up pattern [14]. The warehouse design includes the following stages: (i) data normalization, and (ii) data mart generation.

The *Extract-Transform-Load* (ETL) process is an integral concept in data warehousing. The process involves the extracting, transforming and loading of source systems' data (operational, transactional or legacy databases) into a designated database. In data warehouse systems data can be extracted from either internal or external sources, but in most cases, both [4]. These data need to be cleaned and transformed to provide fast response time and promote the organization's information agility. For most data warehouse assignments, the ETL process usually accounts for up to 70% of the project's timeline [15]. The level of source systems, the nature of data cleaning and transformation varies from basic to complex.

Metadata is "structured information that defines, explains, locates, or otherwise makes it easier to retrieve, use, and/or manage an information resource" [5]. Metadata enables information to make sense of such data as documents (e.g. datasets, images etc.), concepts (e.g. classification schemes), and real-world objects (e.g. organizations, places, paintings etc.).

Within the context of educational resources, metadata are descriptive. This deals directly with the description of the learning resource for discovery/identification purposes.

For OER, the difference between descriptive and administrative metadata (i.e. technical information required to manage a collection of resources) is essential.

There three stages of metadata lifecycle are:

(i) collection,
(ii) maintenance,
(iii) deployment

Web and Metadata

The most widespread languages for metadata description/management are:

- **Resource Description Framework (RDF)**: This is a graphical notation to combine, exchange and share well-structured and semi-structured information [16, 17]

- **XML/Tree**: This is a markup language that outlines an array of rules for coding documents in formats readable to both human and machine [17]. XML documents are normally used for the exchange of data. XML documents comprise markup, which is a type of metadata. Markup is text combined with the resource content

The essence of metadata in OER is to prepare the foundation for learning object agility in terms of (re)usability, discoverability, and interoperability. The standards/approaches existing are:

- Dublin Core: This schema contains a relatively small set of vocabulary terms, describing Web resources (e.g., video, Web pages, images, etc.) [18]
- Learning Object Metadata (LOM): This standard, based on IEEE specification, stipulates the syntax and semantics

Automatic Generation of Metadata

This approach includes: (i) extraction, and (ii) harvesting [19].

The first aspect deals with the resource content. The second one, metadata harvesting, is principally concerned with a practice used in gathering metadata automatically from separate repositories in which metadata have been previously created by either a digital or manual approach. Harvesting is based on the following standard formats and protocols: OAI-PMH, MARC XML, and MODS, to name a few [19–22].

Content extraction involves various computing techniques used for extracting data from information resources, such as Kea and TF.IDF [23, 24].

Similarly, automatic indexing uses machine learning algorithms and/or rule-based methods to extract values [25].

Both of the two above methods, content extraction and automatic indexing, rely on text/data mining for (automatic) extraction of metadata elements.

The Metadata Warehouse Architecture

The core of this architecture is a Web application comprising data source, staging area (ETL process), metadata repository, and a portal (Fig. 4.3).

The ETL process (Fig. 4.4) is a database pattern for performing data integration in warehouses and repositories. The implementation principle of the ETL proposed is hinged on the rule-based/template approach. The rule-based model is an approach that uses programmed instructions to specify how data should be extracted from targeted sources.

The inputs from the source databases came into our system in HTML format. These HTML elements were extracted from the major existing e-learning systems such as OER Commons, edX, Coursera, etc. The extraction happened in two phases, namely: meta-tag extraction and content extraction. We used meta-tag extraction to find out whether the source systems contained the information relevant to the major OER-related keywords.

Data outputs from the DOM locator were parsed to extract the data specified in information fields. This was done by means of XPath extraction rules and regular expressions. These rules were written to manage certain fields containing the target data. We selected each node by moving along or following a path. We used appropriate modes with nodes for each root, element, attribute and text fragment.

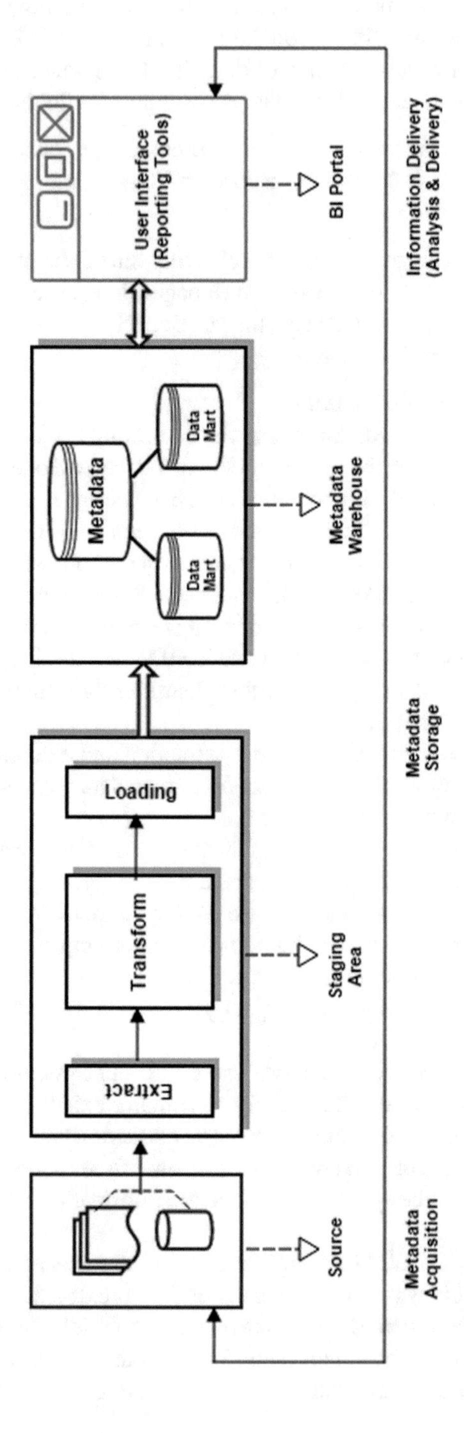

Fig. 4.3 OER metadata architecture

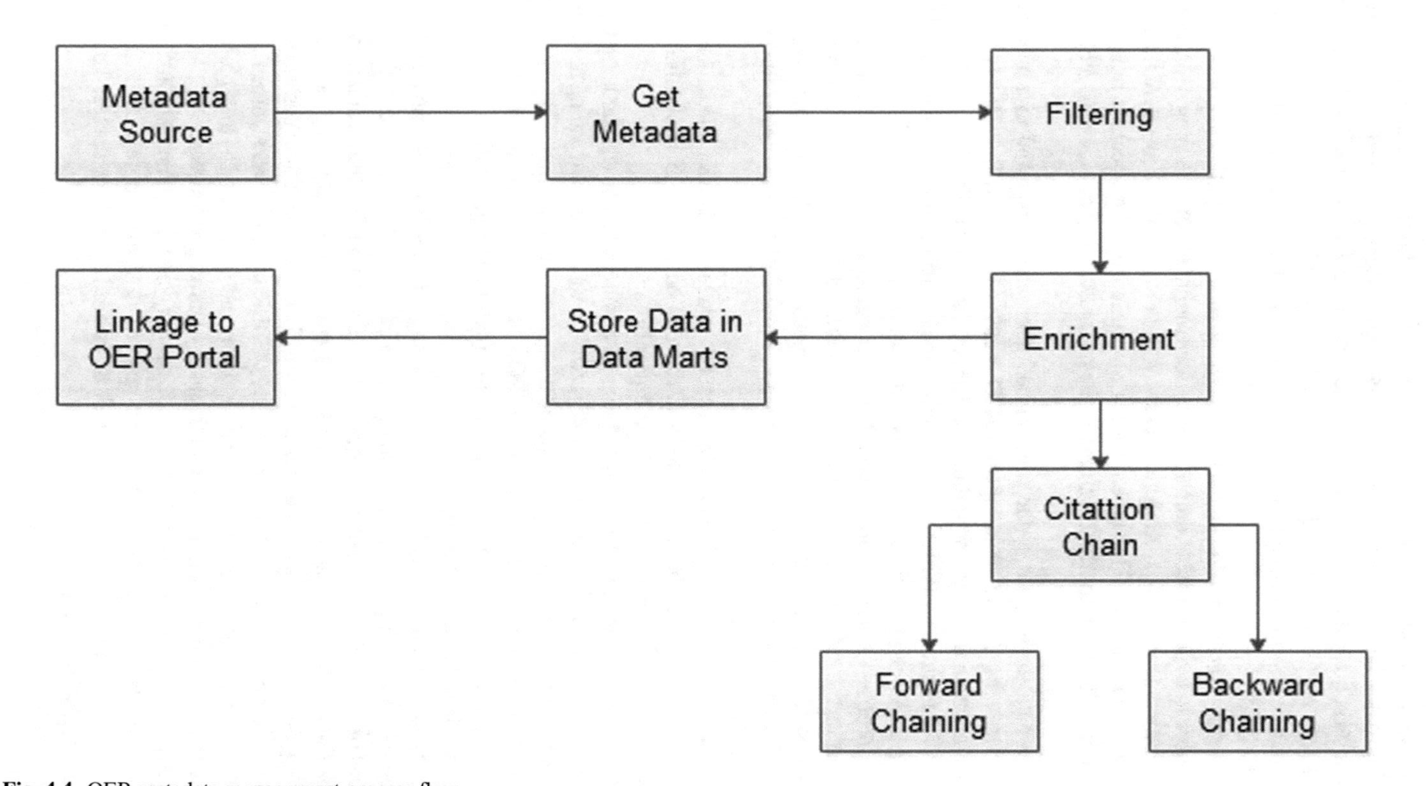

Fig. 4.4 OER metadata management process flow

Our pattern-based approach to data sources made XPath an effective way for capturing data. XPath-based extraction rules were essentially used in processing information sources that provided a Web-based interface. During runtime, a specified Web page was parsed and the HTML DOM was formed. Further, extraction rules, coded in XPath, were applied to the target DOM element to extract the specified metadata fields.

At the data transformation stage, each block of rules was applied to convert data from source to target format. Our transformations included data cleaning, validation, reformatting, aggregation, sorting etc. These tasks were essential because of the need to reduce irrelevant data. This latter process was quite important because noisy data could undermine the important features in the metadata and limit their usability, i.e. agility.

At the data loading stage, the transformed data was transported to a specified back-end system, the metadata repository. This task was done with a series of SQL statements which directly stored data in different data marts. The repository stored metadata for future use. With this goal in mind, the repository was designed to reflect the key features of educational materials and resources.

The purpose of the metadata portal was to aggregate and analyze metadata deposited in the different data marts based on the user's request. Each data mart was dedicated to a particular segment of the metadata warehouse. Upon request, the portal integrated the corresponding metadata and delivered the result through a user-friendly interface.

Experimentation included integrating a set of Coursera and edX course metadata. One subset was based on the author's MOOCs developed in 2015–2016 and discussed in more detail in [26]; this worked well as a proof of concept.

Our case study centered around designing an architecture for warehousing OER metadata. To gain more insight about these methods, we surveyed trending metadata management tools for educational documents and related domains. They included Apache Tika (meta-tag harvesting), Apache POI-Text Extractor (content extraction), Omeka (automatic indexing), Data Fountains (automated generation of extrinsic data), and DSpace (social tagging).

The system architecture tilted toward achieving a better level of discoverability of Web-based learning materials by adopting existing well-researched and proven methods of metadata management. Therefore, in designing the architecture and the process flow, certain management methods and approaches for providing metadata on internet platforms were taken into consideration. The results of the experimentation showed that the architecture improved agility of OER metadata management.

Our next steps are: (i) to move beyond capturing metadata, and (ii) to explore how to raise the agility bar of discovering high quality learning resources.

To achieve this, concepts such as elastic search, citation chain, and automatic indexing will be tinkered with. The combined effect of these computational techniques is expected to help create a network of useful learning resources. This will form the basis for future work.

4.4 Aircraft Communication System

In 1970, the *Aircraft Communications Addressing and Reporting System* (ACARS) was developed to provide up-to-date data on any civil aircraft status. However, increasing technical complexity of avionics and aviation systems complicated analysis of these data.

Literally every key node of an aircraft is connected to the network, including engines, flaps and landing gear. In total, these nodes generate about 500 Gb of data in a single flight; some of these data are transmitted to the base in real time.

In this section, we propose a distributed architecture of the flight data analyzer. The analyzer uses a classic client-server pattern; the client is a Web interface. The server emulates the flight data aggregator; it provides data for each flight.

ACARS is the electronic messaging system similar to SMS in the format. In the flight, the aircraft transmits messages. Stations receiving these messages are all around the globe; they receive these messages and forward them to a certain airline. An airline or dispatching center may transmit the data to the station prior to forwarding them to a specific board. Data examples are: the route, aircraft load, balance, amount of fuel. ACARS can transmit detailed data on avionics and aircraft systems, e.g. an inertial navigation system and a Doppler accelerator [27].

ACARS also transmits detailed data concerning failures; however, this data can be misleading and therefore result in a crisis [28]. Once, due to sensors damaged by engine explosion, ACARS reported failure of all aircraft systems. The professionalism of the pilot, i.e., a crisis-responsive human factor, avoided catastrophe. One of the reasons for the crisis was that the experts analyzed a large amount of contradictory textual data [29]; that is the reason to create an analyzer with graphical output.

ACARS includes aircraft equipment and an extensive ground part [30]. The data transfer rate is 2400 baud. To receive the ACARS signal on VHF, a normal scanner or receiver, tuned to the frequency of the band, is required.

The airborne aircraft system includes a management unit, which receives/transmits messages, and a control unit, which interacts with the crew and displays messages. The ground part includes a network of transceiver stations and computer systems. The system provides two-way airline-to-aircraft messaging during the flight. The messages are either downlinks (from aircraft to ground) or uplinks (opposite). There is no uniform format as each airline uses a proprietary format. Depending on the airline, the system determines message format.

Applications for ACARS data support one of the two modes: (i) import historical logs for data analyzing and reporting; (ii) get live data from real-time Web services (e.g. SkySpy, AirMaster 2000 etc.). The most common ACARS data analyzers are: Cosmic ACARS Analyzer and Pervisell ACARS Log Analyzer.

Table 4.1 represents advantages and disadvantages of the existing solutions. Both are desktop applications. They do not support rich/thick client mode. The applications are not connected to central server; their instances have no common database, and therefore the system is not data consistent. Thus, we recommend migrating the data analyzer to Web environment and using a Web service as a pattern. As we will see, this improves agility and fits into the context of the flight data analyzer.

ACARS classifies messages by their content as follows: (i) Air Traffic Control (ATC); (ii) Aeronautical Operational Control (AOC), and (iii) Airline Administrative Control (AAC) [31].

AAC format depends on the airline: e.g., Russian S7 Airlines has a proprietary message format [30, 32]. Therefore, the analysis of this type of messages is out of our scope.

AOC serves the communication applications between the aircraft and the airline (or the airline's service partner) [33]. AOC messages are user-defined, and there are solutions for the ground service systems (e.g. ACARS Service Management by Sabre Airlines Inc.). Since such systems should be compatible with the ground service control, this type of messages is out of our scope.

The ATC messages include OOOI, METAR, and POS messages [30].

The OOOI type is for tracking the aircraft and flight status, monitoring the crew parameters etc. Examples of the data include: pushback from a gate, wheels off, wheels on, number of passengers, wheel chairs required, arrival time and gate etc.

At the start of each flight phase, the aircraft transmits these messages to the airline and ground services:

(1) METAR type is for communicating between the Aviation Weather Center (AWC) and the aircraft. Weather data is out of our scope as this data is stored in AWC servers and they analyze it as required
(2) POS messages that indicate aircraft position and include data on the registration number of the aircraft, flight suffix, current height, longitude and latitude, and flight number

The product vision is presented in Table 4.2.

The input data flows are the ACARS data from the ACARSD online server, and from the log files of SkySpy, AirNav and JACARS. The output data flows include: graphics/table online reports, PDF, Excel and CSV formats.

ACARS data analyzer connects to the data suppliers; it stores non-parsed data. The data collector analyzes the structure of the message and determines the message type. Based on message type, ACARS data collector determines the aircraft type in order to determine the message structure. After the message structure is determined, ACARS processor starts parsing it. In case of success, the collector writes the data into the table.

The Monitoring Web Application starts monitoring the table/structure with its controller, which updates the view structure. This module is a part of the Web interface, it displays data, configures system timeouts and updates, data suppliers etc. This module displays historical flight data; it contains system controllers.

The logical view of the system is in Fig. 4.5.

Table 4.1 Advantages and disadvantages of the existing solutions

Solution	Advantages	Disadvantages
Cosmic ® ACARS Analyzer	Real time connection to SkySpy, AirNav, JACARS etc.	Alerts are working only in online mode
	Data alerts for registrations, flight numbers, message labels and message contents	Online mode is not compatible with data analyze mode, including parsing the ACARS messages
	Option for logging empty data messages	All reports are available in HTML only
	Import 3rd party log file from WACARS, AIR 2 K, SkySpy Logs, SkySpy Database, AirNav Logs WinRadio and JACARS	Data stored in single database for a given desktop application (no shared database for different instances of application)
	Customizable Airline/Aircraft/Routes/Flight Lists	Real time data collection only up to one month due to large data volume
	Statistical information available—Top 10 Aircraft, Flight Number, Airlines, Routes and Types as well as general message statistics	Correlation of the data imported in real time and the data imported from log file is not clear (should we update the real time data by the log data?)
	–	No graphical reports, only tables
	–	No filter for incoming data. Filter is critical for the airline, the key stakeholder, there is no need to store and analyze ACARS messages of 'the other' airlines
Pervisell ® ACARS Log Analyzer	A lightweight solution for the monitoring messages on the fly	Real time data are stored in temporary database and lost at logout
	Real time connection to SkySpy, AirNav, JACARS etc.	No graphical reports, only tables
	The database includes look-up tables for Aircraft and Carriers	Online mode in not compatible with data analyze mode, including parsing the ACARS messages
	There are additional databases for Routes and Message Labels	Alerts are not implemented
	Exporting data in HTML, Excel formats	No filter for incoming data. Filter is critical for the airline, the key stakeholder, there is no need to store and analyze ACARS messages of 'the other' airlines
	–	The log files data (imported from CSV) are stored in temporary database and lost at logout
	–	Reports are not available 'online': to build them, user must disconnect the real-time data storage
	–	Desktop application with data, which is not stored in the database during the sessions
	–	Export to XML, CSV formats is not implemented

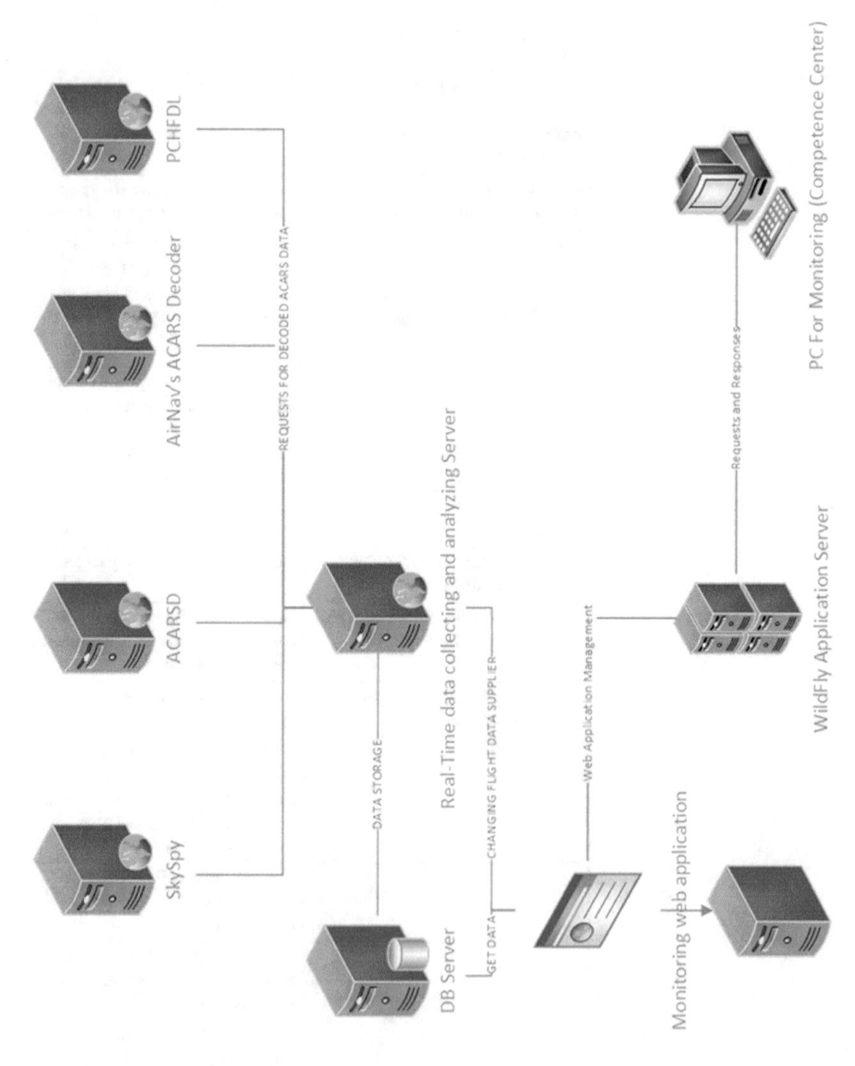

Fig. 4.5 Logical view of the information system

Table 4.2 Product vision

Criteria	Description
For	For the representatives of security and flight monitoring departments of airlines as well as aviation technical experts
Who	Need to analyze large amounts of historical flight data, being at the same time able to receive notifications about critical and important flight events Need to display data in a convenient graphical form Need to export data in various data formats, including printed reports Need a centralized data analysis system with a single and unique database for all experts Need long-term storage of flight data Need data synchronization with centralized flight data provider servers, with the ability to manage data federation
The	ViOne ACARS Online Analyzer
That	Analyzes flight data online using the ACARS messaging system
Unlike	Cosmic ® ACARS Analyzer, Pervisell ® ACARS Log Analyzer
Our product	Provides customizable notifications of critical flight events Performs an analysis of the flight data for a specific board for any period of time Allows to a user to create printed reports in PDF Outputs analytical information in tabular and graphical form Allows export of flight data to common formats Allows import of ACARS data from open source to PDF (including replaces and updates)

The system can connect only to the ACARSD Web service as ACARSD is an open-source server, which provides data of the same format as the commercial services. We designed the application pattern in an agile way to promote modifiability.

Therefore, replacing ACARSD with a commercial Web server requires only a quick configuration change in the application settings.

A Web application with a Web service module was deployed using Maven, and Application Server WildFly with the JSP container. Configuration Web server and DBMS server were deployed on the same machine. The system communicates with the DBMS using the JDBC based Hibernate Framework. The client browser communicates with the deployed application via HTTP protocol.

Deployment view is presented in Fig. 4.6.

The system layer view is in Fig. 4.7.

For better agility, the system pattern is layered. This promotes modifiability by means of global interfaces [34, 35].

The presentation layer contains user interface components. The Web pages based on these components are stored in the same layer. These pages contain no logic, so this layer has a GUI controller i.e. the interaction point with the other layers. Our GUI used a pattern-based solution including forms, scripts and markup. Therefore, this layer contains the Primefaces Library.

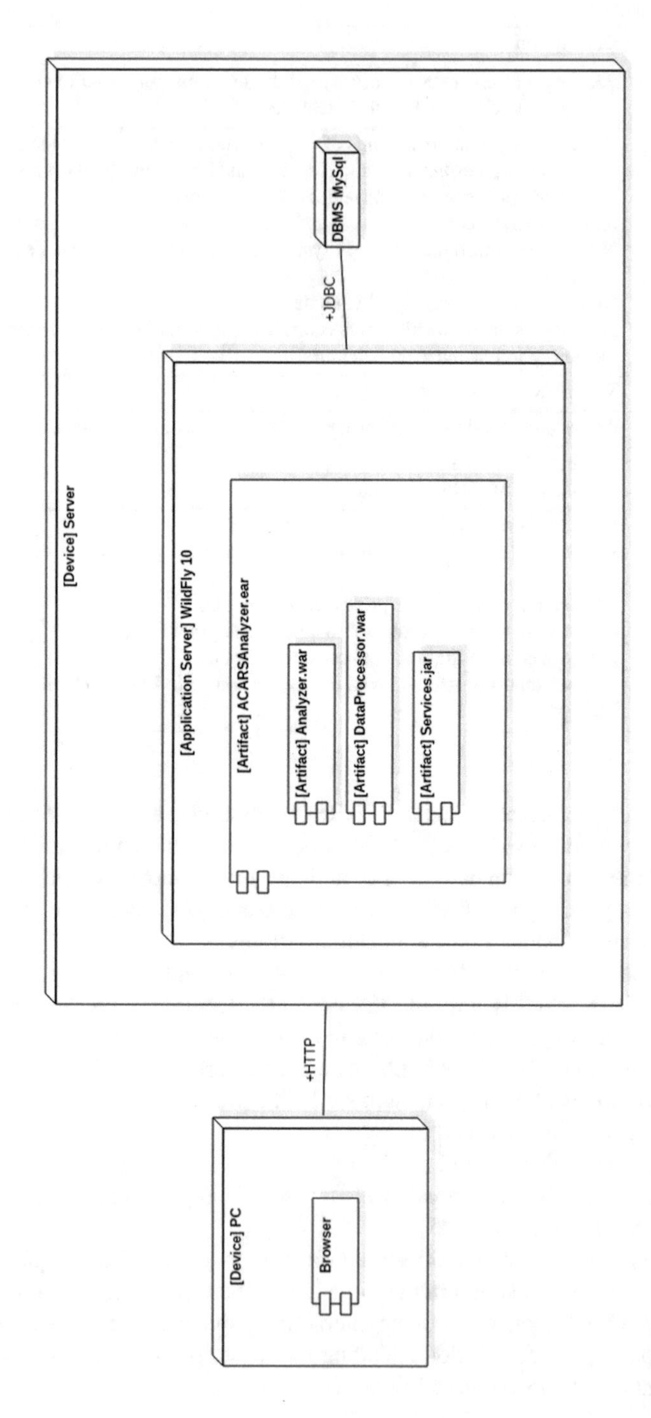

Fig. 4.6 Deployment view

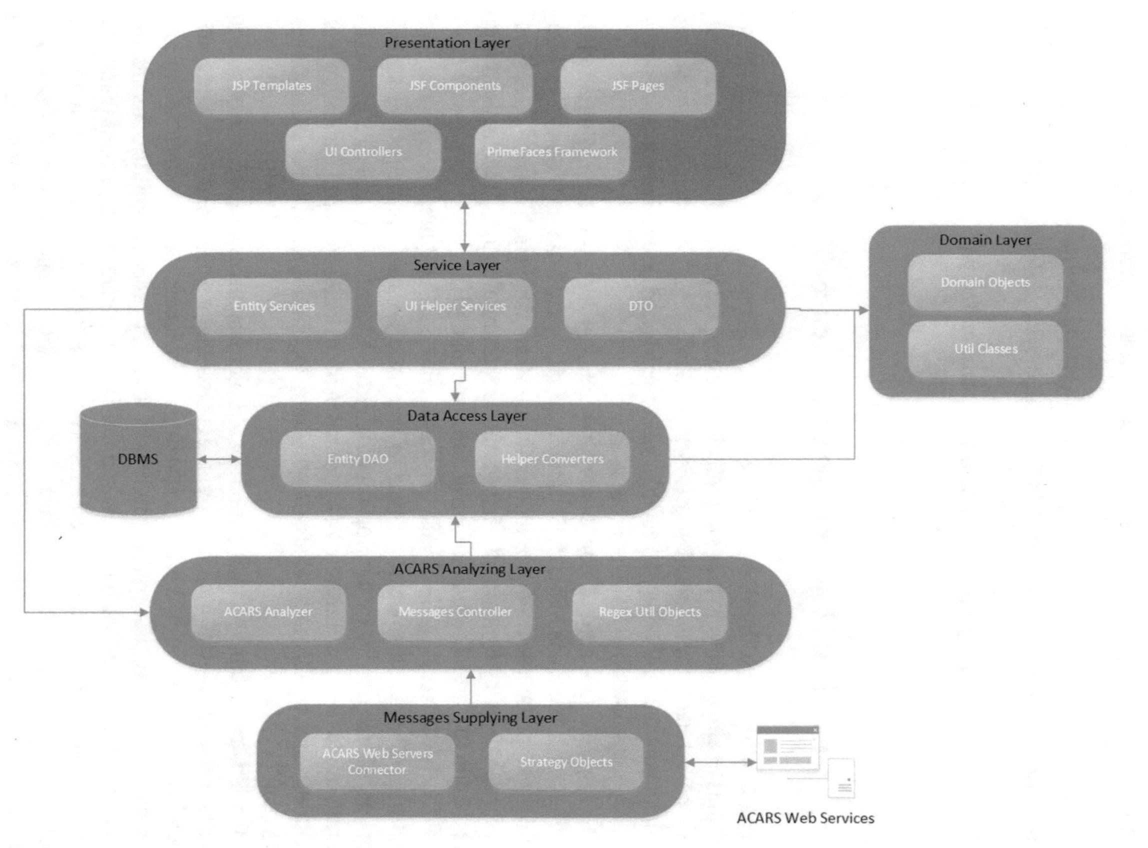

Fig. 4.7 Layers view

The service layer is also a GUI part of the system. The services operate with the presentation view via the other layers.

The domain layer contains object mapping entities and their utilities.

The data access layer contains data access objects [36] for database querying. The DBMS layer represents the database; it contains converters for object mapping interfaces [26].

The ACARS analyzing layer is the core of the system. This layer contains components designed to process incoming ACARS messages: analyzer, message processing controller, and regular expression handlers.

The message supplying layer contains connectors for Web servers and ACARS message servers.

4.5 Conclusion: How Patterns and Practices Work

For agile software product development, efficient communication between the client and the developer sides is mission-critical. This chapter outlined patterns and practices for resonant communication between these two sides. Our approach combined models, methods and computer-aided tools to tailor and customize software development and knowledge transfer processes.

To improve this kind of communication, our key idea was adding an efficient bi-directional feedback to the basic patterns of large-scale and mission-critical software development process presented in the previous chapters. This initial idea was further enhanced by dynamically adjustable "signal amplification" mechanisms with an addition of "useful resonance" generator (see also [2]).

Further, our resonant feedback generator with carefully tuned monitoring and adjusting mechanisms became a part of the pattern-based metacognitive knowledge transfer and software development lifecycle. As such, we elaborated our initial pattern-based idea and illustrated it by means of human factor-related features, which included metacognitive knowledge acquisition, negotiations, and other communicative "soft" skills.

We illustrated this resonant pattern-based metacognitive software development and knowledge transfer approach by a few case studies. The first case study was knowledge transfer in a diverse university environment, where transmitting and receiving sides were essentially different in terms of process maturity, culture and mentality. The second case study was a uniform metadata management pattern to handle diverse information resources for worldwide knowledge transfer in massive online open education. The third case study was a large-scale software development project that improved communication (i.e. knowledge transfer) in a mission-critical aircraft control system.

For further reading regarding the enterprise-scale software quality products efficiently implemented, we would like to recommend such sources as: [2, 37]. These references contain case studies on the enterprise systems for strategic resource management in oil-and-gas production and distribution, air traffic management, and nuclear power generation. Our implementations incorporated DSL-based layered patterns and CASE tools, which essentially simplified the entire software development lifecycle, particularly such processes as product analysis, design and re-engineering. To further enhance agility, we recommend cloning these high level patterns and then tailoring these clones according to the requirements of each particular client.

References

1. Gamma, E., Helm, R., Johnson, R. & Vlissides, J. (1994). *Design patterns: Elements of reusable object-oriented software* (1st ed.). Addison-Wesley.
2. Zykov, S. (2016) *Crisis management for software development and knowledge transfer issue 61: springer series in smart innovation, systems and technologies* (133 pp.). Switzerland: Springer International Publishing.
3. Kuchins, A. C., Beavin, A., & Bryndza, A. (2008). *Russia's 2020 strategic economic goals and the role of international integration*. Washington, D.C.: Center for Strategic and International Studies.
4. Majid, A., Larissa Moss, T. & Sid, A. (2005). Data strategy.
5. Riley, J. (2017). Understanding metadata: What is metadata, and what is it for? NISO Press.
6. Geith, C., & Vignare, K. (2008). Access to education with online learning and open education resources: Can they close the gap? *Journal of Asynchronous Learning Networks, 12*(1), 1–22.
7. Belawati, Tian. (2014). Open Education, Open Education Resources, and Massive Open Online Courses. *International Journal of Continuing Education and Lifelong Learning, 7*(1), 2014.
8. D'Antoni, S. & Catriona, S. (2009). *Open Educational Resources: Conversations in Cyberspace*. UNESCO Publishing.
9. Ischinger, B. (2007). Giving knowledge for free: The emergence of open educational resources. Centre for Educational Research and Innovation.
10. Hylén, J. (2006). Open educational resources: Opportunities and challenges. OECD's Centre for Educational Research and Innovation, Paris, France. www.oecd.org/edu/ceri.
11. Krause, S.D. & Lowe, C. (2014). Invasion of the MOOCs: The promises and perils of massive open online courses parlor press anderson. South Carolina.
12. Gray, P., & Waston, H. J. (1998). Present and future directions in data warehousing. *Database for Advances in Information Systems, 29,* 83–90.
13. Turban, E. & Aronson, J.E. (2001). *Decision support systems and intelligent systems* (6th ed.), Upper Saddle River, NJ: Prentice Hall (Copyright 2001).
14. Ţole, A.A. (2015). The importance of data warehouses in the development of computerized decision support solutions. A comparison between data warehouses and data marts. *Database Systems Journal*, *VI*(4).
15. Kimball, R. & Caserta, J. (2004). The Data Warehouse ETL toolkit practical techniques for extracting, cleaning, conforming, and delivering data. Wiley Publishing, Inc.
16. Patel-Schneider P.F., & Simeon, J. Building the semantic web on XML.
17. Decker, S. & Melnik, S. The semantic web: The roles of XML and RDF.
18. Ebner, H. *Introduction to Dublin core metadata*. Sweden: Knowledge Management Research Group. Royal Institute of Technology (KTH).
19. Greenberg, J., Spurgin, K. & Abe Crystal, Final Report for the AMeGA (AutoZmatic).
20. Broughton, V., Polfreman, M., & Wilson, A. Metadata generation for resource discovery.

21. Park, J., & Lu, C., Application of semi-automatic metadata generation in libraries.
22. Peake, M. (2012). Open archives initiative protocol for metadata harvesting, dublin core and accessibility in the OAIster Repository. Library Philosophy and Practice (e-journal).
23. Kea Automatic Keyphrase Extraction homepage. http://www.nzdl.org/Kea/index_old.html.
24. http://stevenloria.com/finding-important-words-in-a-document-using-tf-idf/.
25. Mark Patton et al. (2004) Toward a metadata generation framework: A case study at Johns Hopkins University. *D-Lib Mag*, *10*, 11.
26. Data Transfer Object. Electronic resource. https://en.wikipedia.org/wiki/Data_transfer_object.
27. Ed Flynn (1995). *Understanding ACARS* (3rd ed.).
28. Incident with the Airbus A380 over the island of Batam. Electronic resource. https://ru.wikipedia.org/wiki/A380_Batam.
29. PlanePlotter, ACARS & Boeing-787 Dreamliner Emergency. Electronic resource. http://adsbradar.ru/787/planeplotter-acars-i-boeing-787-dream-emegcy.
30. Kucheryavii, A. (2016) Avionics. Lani.
31. ARINC 619. (1994). ACARS protocols for avionic end systems.
32. Okan, D. Fly safe. Electronic resource. http://denokan.livejournal.com/.
33. Aeronautical Operational Control. Electronic resource. https://en.wikipedia.org/wiki/Aeronautical_Operational_Control.
34. Fowler, M. (2016). *Patterns of enterprise application architecture*. Moscow.
35. Gamma, E. (2016). *Design patterns*. Piter.
36. Data Access Object. Electronic resource. https://en.wikipedia.org/wiki/Data_access_object.
37. Zykov, S. (2016). The online evolution: From early repositories to state-of-the-art MOOCs. In: *Central and Eastern European Software Engineering Conference in Russia (CEE-SECR '16), Moscow*, Russian Federation 2016, Article No.: 9. New York, NY, USA: ACM.

Conclusion
Agility Revisited: What, Why and How

As we have seen, crisis management typically requires agility. We can interpret crisis management as a triangle, the sides of which are the global project optimization parameters: processes, resources, and constraints—see Fig. 1. The inner triangle implies a *comfort zone*; in this case the global parameters are well balanced so that project management requires no attention. The outer triangle implies a *compromise zone*; in this case the global parameters require agile balancing so that the project becomes manageable. The zone outside the outer triangle is the *crisis zone*; in this case the global parameters (i.e. predetermined/approved product requirements/constraints) require agile readjustment so that the project becomes manageable. We can also refer to the first zone as a green one, to the second zone as a "local" crisis or yellow one, and the third one as a "global" crisis or red one.

To clarify the above diagram, let us provide a brief explanation and some examples. Each of the optimization dimensions includes multiple subdomains. For instance, processes typically include roles and artifacts. Constraints typically include business and technological ones. Resources typically include HR, budget, and time. These subdomains are crisis management parameters. In a local crisis, a single parameter usually needs adjustment; so we can compromise on the others to reach the agility balance required. For instance, for a reduced budget we can ask for more time to finalize the product, or produce mission-critical functions only within the timeframe initially approved.

A global crisis is more complex as a number of dependent parameters requires readjustment; and we cannot reach this under the initial requirements/constraints. Thus, we have to backtrack and revise/adjust these requirements/constraints (one by one, depending on their criticality), and then proceed with the new readjusted triangle. Constraint examples are: system response time, network throughput, maximum number of concurrent users, and database size. Thus, we can either decrease the maximum number of concurrent users or increase system latency to adjust the global parameters, and then start over with the updated processes and resources if necessary. To adjust the processes, we can either reduce the number of artifacts or simplify some of the artifacts (e.g. substitute a detailed requirement

S. V. Zykov, *Managing Software Crisis: A Smart Way to Enterprise Agility*,
Smart Innovation, Systems and Technologies 92,
https://doi.org/10.1007/978-3-319-77917-1

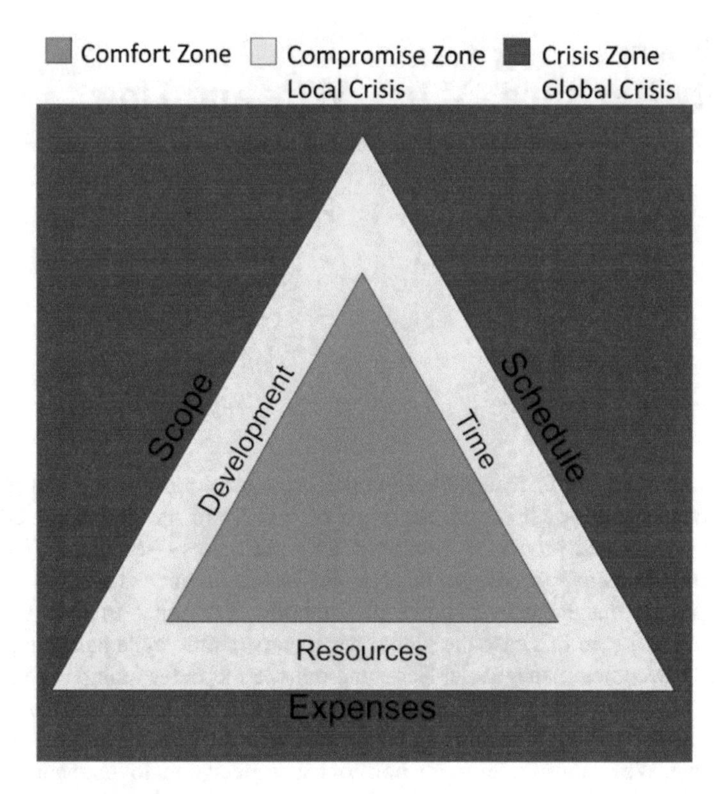

Fig. A.1 Crisis management triangle

specification document with a brief requirement checklist). Another option for process adjustment would be reducing the number of roles or combining certain roles, e.g. assigning: (i) project and product manager or (ii) chief architect and (iii) detailed designer roles to the same individual.

Concerning the color zones in the crisis optimization/management diagram, green does not necessarily mean being careless; optimizing resources in this situation may help managing concurrent projects running low due to a local or global crisis. Yellow typically means that careful monitoring and project parameter adjustment is required; otherwise a local crisis may become global. Red does not necessarily mean hopeless or helpless; however, it means that immediate attention is required as the initial plan will not work in the new crisis circumstances.

Keeping in mind this diagram together with our initial Enterprise Agility Matrix (Fig. 1.1) and the continuous cognitive "monitor-analyze-adjust" software production cycle will guide and guard the development processes and make them crisis resistant due to constantly improving agility.

One more takeaway on crisis management is the following Tightrope Walker metaphor. Imagine a person walking on the rope high above the ground. The Tightrope Walker holds a balancer in his hands. The clock is ticking: if he reaches the destination too late, he loses. The balancer looks helpful, and it is sometimes. However, every now and then (e.g. due to the side wind) it happens to be unstable. In this case, the Tightrope Walker changes his grip to stabilize the balancer. If this is not enough, the Tightrope Walker changes position (which resembles a dance figure) to stabilize the balancer. A side wind may result in the harmful resonance of the rope which requires complicated actions (up to a complete backtrack and restart) to survive.

Why did we use this metaphor? This is a model of a crisis. The Tightrope Walker is the crisis manager. The rope is the development process with a product concept as the starting point and the product delivery (or retirement, depending on the preferences) as the end of the way. Late product delivery is often risky or fatal. The two edges of the balancer are resources and constraints. Balancer instability means a crisis. Walking means the lifecycle. Better balance means better agility. More specifically, the speed and efficiency of a return to a balanced state are agility (or crisis responsiveness) metrics. Stabilizing the balancer by a grip change means a "local" crisis. Stabilizing the balancer by an extra backtrack (or complete restart) means a (more) "global" crisis.

This problem is somewhat similar to multi-criteria optimization; the difference, however, is that the constraints are subject to uncertain changes.

Another finding is that a crisis is similar to a harmful resonance, and the software developer needs a positively resonant feedback to compensate that negative influence. In case of "global" crisis caused by a harmful resonance, insufficient agility may result either in a complete lifecycle restart or in fatal consequences (the latter is out of the scope of this book).

We sincerely hope that our metaphors and instances (including but not limited to dancing dinosaurs, Enterprise Agility Matrix, and the Tightrope Walker) will provide helpful hints for efficient management and overall product success, even in crisis. We wish you ultimate agility, and good luck with your software product development.

Glossary

Activity A thing that a person or group does or has done, a relatively small isolated task with clear exit criteria

Adaptive maintenance Maintenance, which modifies system in order to adapt the product to the new software and hardware environment

Agile methodology An alternative to traditional project management where emphasis is placed on empowering people to collaborate and make team decisions in addition to continuous planning, continuous testing and continuous integration

Agility Ability to adapt to uncertainties and changes of environment

Anthropic-oriented Relating to human

Architectural heterogeneity A property of a system, which includes components based on different architectural patterns

Architecture Centric Design Method (ACDM) A novel method for software architectural design developed by the Software Engineering Institute at the Carnegie Mellon University

Architecture Tradeoff Analysis Method (ATAM) Risk-mitigation process used early in the software development lifecycle. ATAM was developed by the Software Engineering Institute at the Carnegie Mellon University

Abstract Syntax Tree (AST) A tree representation of the abstract syntactic structure of source code written in a programming language

Backlog An accumulation of uncompleted work or matters needing to be dealt with

Best practice(s) Commercial or professional procedures that are accepted or prescribed as being correct or most effective

S. V. Zykov, *Managing Software Crisis: A Smart Way to Enterprise Agility*,
Smart Innovation, Systems and Technologies 92,
https://doi.org/10.1007/978-3-319-77917-1

Bitmessage A decentralized, encrypted, p2p, trustless communications protocol that can be used by one person to send encrypted messages to another person, or to multiple subscribers

Blockchain A continuously growing list of records, called blocks, which are linked and secured using cryptography; each block typically contains a hash pointer as a link to a previous block; it is impossible to change the content of a block without affecting every subsequent block

Bloom's taxonomy Bloom's classification of levels of intellectual behavior important in knowledge transfer

Business architecture Defines the structure of the enterprise in terms of its governance structure, business processes, and business information

Business Process Management (BPM) A discipline in operations management that uses various methods to discover, model, analyze, measure, improve, optimize, and automate business processes

Build-and-fix, model A model of software development without a deliberate strategy or methodology. Programmers immediately begin coding. Often late inthe development cycle, testing begins, and the unavoidable bugs must then be fixed before the product delivery

Business requirements Critical activities of an enterprise that must be performed to meet the organizational objective(s) while remaining solution independent

Coinbase transaction The transaction inside a block that pays the miner his block reward

Computer-Aided Software Engineering (CASE) tools A domain of software tools used to design and implement applications

Common Interface Definition Language (CIDL) A DSL based on Lua that includes communication types, interface definition functions, type definition functions and Lua syntax support

Carnegie Mellon University (CMU) A private research university in Pittsburgh, Pennsylvania

Cohesion Degree to which the elements of a certain module belong together

Common vision Essential component of a learning organization because it provides the focus and energy for learning; a realistic, credible, attractive future for an organization Common vision is derived from the members of the organization, creating common interests and a sense of shared purpose for all organizational activities

Compatibility of GUI A function the user interface performs with users' expectation of the system; GUI should match user's prior knowledge of the system software and hardware

Computer-Aided Design (CAD) Software used by architects, engineers, drafters, artists, and others to create precision drawings or technical illustrations. CAD software can be used to create two-dimensional (2D) drawings or three-dimensional (3D) models

Computer-Aided Engineering (CAE) Broad usage of computer software to aid in engineering analysis tasks

Computer-Aided Manufacturing (CAM) An application technology that uses computer software and machinery to facilitate and automate manufacturing processes

Consistent GUI Quality of interface that ensures that user can easily understand how it works in general thus ensuring smooth transition between different areas of the GUI

Corpus-based technique A technique to determine text sentiment by finding the co-occurrence of patterns of words

Corrective maintenance Maintenance, which fixes existing defects in the software product without changing the design specifications

Commercial Off-The-Shelf (COTS) Satisfy the needs of the purchasing organization, without the need to commission custom-made, or bespoke, solutions

Coupling degree To which one component knows about the inner workings/elements of another one

Courage The ability to do something that frightens one; bravery

Course climate A set of overarching and pervasive values, norms, relationships, and policies of the course that make its character

Crisis Misbalanced production and realization of a surplus value, the root cause of which is separation between the producers and the means of production

Customer Relationship Management (CRM) Practices, strategies and technologies that companies use to manage and analyze customer interactions and data throughout the customer lifecycle, with the goal of improving business relationships with customers, assisting in customer retention and driving sales growth

Decentralized Application (Dapp) Application that runs on a P2P network of computers rather than a single computer

Data warehouse A large store of data accumulated from a wide range of sources within a company and used to guide management decisions

Database Management System (DBMS) Software that handles the storage, retrieval, and updating of data in a computer system

Deliverable A practical outcome, any measurable project artifact, which is a result of each project task, work item or activity

Design pattern A general reusable solution to a commonly occurring problem within a given context in software design

Design specification A detailed document providing information about the characteristics of a project to set criteria the developers will need to meet

Design A software development lifecycle stage, which formally describes the components of the future software product and the connections between these components

Double spending A potential flaw in a digital cash system, resulting in successfully spending some money more than once

Domain Specific Language (DSL) A computer language specialized to a particular application domain

Domain Specific Modeling (DSM) A software engineering methodology for designing and developing systems, such as computer software

Enterprise Architecture (EA) A discipline for proactively and holistically leading enterprise responses to disruptive forces by identifying and analyzing the execution of change toward desired business vision and outcomes. EA delivers value by presenting business and IT leaders with signature-ready recommendations for adjusting policies and projects to achieve target business outcomes that capitalize on relevant business disruptions

Enterprise Content Management (ECM) Formalized means of organizing and storing an organization's documents, and other content that relate to the organization's processes. Encompasses strategies, methods and tools used throughout the lifecycle of the content

Enterprise engineering matrix Matrix, the columns of which correspond to processes, data and systems, and the rows of which contain enterprise system levels. Used to detect and predict local crises of software production

Enterprise Resource Planning (ERP) Business process management software that allows an organization to use a system of integrated applications to manage the business and automate many back office functions related to technology, services and human resources

Education Quality Control System (EQCS) A system used to control education quality

Ethereum A decentralized platform that runs smart contracts: applications that run exactly as programmed without any possibility of downtime, censorship, fraud or third party interference

Ethereum message Data (as a set of bytes) and Value (specified as Ether) that is passed between two Accounts, either through the deterministic operation of an Autonomous Object or the cryptographically secure signature of the Transaction; it contains five fields: a sender address, a recipient address, an ether amount, a data field, and a startgas value

Ethereum transaction A piece of data, signed by an External Actor. It represents either a Message or a new Autonomous Object. Transactions are recorded into each block of the blockchain; it contains six fields: a recipient address, an ether amount to be sent, a data field, a startgas value, and a gasprice

Evolutionary, model An iterative model of software development based on the idea of rapidly developing an initial software implementation from very abstract specifications and modifying this according to appraisal

Expectancy The feeling that a person has when he or she is expecting something

Extreme programming, methodology (XP) A software development methodology, which is intended to improve software quality and responsiveness to changing customer requirements; a pragmatic approach to program development that emphasizes business results first and takes an incremental approach to building the product through continuous testing and revision

Far knowledge transfer Knowledge transfer that allows a multi-domain knowledge application in different contexts, some of which are substantially far from the original one

Feedback A helpful information or criticism that is given to someone to say what can be done to improve performance

Formal methodology A codified set of practices (sometimes accompanied by training materials, formal educational programs, worksheets, and diagramming tools) that may be repeatedly carried out to produce software

"Fragile" base classes A problem of object-oriented programming where the superclasses contain seemingly safe modifications, which, when inherited by the derived classes, may cause malfunctions of these derived classes

Heterogeneity A property of a set, which consists of elements that are essentially different from each other

Human-Related factor A factor originating from human nature, which influences requirements elicitation, and, consequently, software development

Implementation A software development lifecycle stage, which produces the code of each individual component of the software product

Incremental, model A model of software development where the product is designed, implemented and tested incrementally (a little more is added each time) until the product is finished. It involves both development and maintenance

In-house development Software development by a corporate entity for purpose of using it within the organization

Integration A software development lifecycle stage, which produces the entire software product out of the individual components implemented previously

Interrelatedness The ability of one system component change to significantly influence a number of adjacent components changes

Interview A valuable source of information that allows to express themselves; it tends to be time consuming and expensive

Internet of Things (IoT) The interconnection via the Internet of computing devices embedded in everyday objects, enabling them to send and receive data

The K-nearest neighbour algorithm An algorithm that stores all available cases and classifies new cases based on a similarity measure, for instance, distance functions

Key Performance Indicator (KPI) Business metric used to evaluate factors that are crucial to the success of an organization

Knowledge Transfer (KT) The practical problem of transferring knowledge from one part of the organization to another

Legacy software product An old software product, of, relating to, or being a previous or outdated computer system. Often implies that the system is out of date or in need of replacement

Lexicon-based approach Approach used in sentiment analysis that relies on lexical resources such as Sentiwordnet, MPQA-wordnet, WordNet-Affect17 etc.; these lexical resources are based on the idea that any text can be assigned a polarity based on the polarity of words in the text

Machine Learning (ML) A field of computer science that gives computers the ability to learn without being explicitly programmed

Maintenance A software development lifecycle stage, which includes all aspects of the product operation and support at the client's site

Massive Open Online Course (MOOC) An online course aimed at unlimited participation and open access via the Web

Mastery A skill that allows doing, using, or understanding something very well

Max entropy A probabilistic classifier based on the Principle of Maximum Entropy (PME) that selects the most uniform distribution (thus maximum entropy) from a set of possible distributions that satisfies the given constraint; Max Entropy is based on the probability that a word sentence or a document belongs to given a context must maximize the entropy of the system

Merkle root A result of recursive pairwise hashing of nodes from branches of the Merkle tree

Merkle tree A data structure used for efficiently summarizing and verifying the integrity of large sets of data; are used in Bitcoin to summarize all the transactions in a block, producing an overall digital fingerprint of the entire set of transactions, providing a very efficient process to verify whether a transaction is included in a block

Metacognitive Something that refers to higher order thinking which involves active control over the cognitive processes engaged in transferring knowledge

Meta-knowledge transfer Practical problem of transferring knowledge about a preselected knowledge

Metaphor A figure of speech in which a word or phrase is applied to an object or action to which it is not literally applicable

Method A particular procedure for accomplishing or approaching something, especially a systematic or established one

Methodology A system of methods used in a particular area of study or activity; a framework that is used to structure, plan and control the process of developing an information system

Milestone A key control point where certain results are achieved; a significant stage or event in the development of something

Modularity The degree to which a system's components may be separated and recombined; uses minimum connectivity between the modules, so that each relatively small and functionally separate task is located in a separate software module

Motivation A force or influence that causes someone to do something

Naïve bayes A simple probabilistic classifier based on applying Bayes' theorem with strong independence assumptions between the features

Near knowledge transfer The knowledge transfer applicable to adjacent problem domains only

Natural Language Processing (NLP) A field of computer science, artificial intelligence and computational linguistics concerned with the interactions between computers and human (natural) languages, and, in particular, concerned with programming computers to fruitfully process large natural language corpora

Nuclear Power Plant (NPP) A thermal power station in which the heat source is a nuclear reactor

Object-oriented, model A model of software development based on object-oriented paradigm

Observability A measure of how well internal states of a system can be inferred from knowledge of its external outputs

Open Educational Resources (OER) Freely accessible, openly licensed text, media, and other digital assets that are useful for teaching, learning, and assessing as well as for research purposes

Online Analytical Processing (OLAP) A category of software tools that provides analysis of data stored in a database. OLAP tools enable users to analyze different dimensions of multi-dimensional data

Online Transaction Processing (OLTP) A class of information systems that facilitate and manage transaction-oriented applications, typically for data entry and retrieval transaction processing

Open vote network A 2-round decentralized voting protocol where all communication is public, and the voter's privacy protection is maximum; the system is self-tallying and dispute-free

Oscillator (LC) circuit An electric circuit, which consists of capacitor and inductive coupling. Uses feedback for oscillation

Perfective maintenance A type of maintenance, which implements changes to the product functional specification, making the new product release with improved functionality and same or better quality in terms of performance, reliability, security, availability, usability etc.

Practice An activity of doing something repeatedly in order to become better at it

Privity A term mostly used in contract law, referring to the connection between parties to a particular transaction

Process A series of actions or steps taken in order to achieve a goal; a sequence of the tasks to be implemented, they are clearly different, i.e. have a clear start and termination criteria, and, in some times dependent on each other

Production Lifecycle Management (PLM) Process of managing the entire lifecycle of a product from inception, through engineering design and manufacture, to service and disposal of manufactured products

Proof-of-Stake (PoS) Type of algorithm by which a cryptocurrency blockchain network aims to achieve distributed consensus, where the creator of the next block is chosen via various combinations of random selection and wealth or age

Proof-of-Work (POW) Process of finding such an integer that when hashed together with the rest of the block header data, the resulting output has at least a given amount of leading zero; this process might be rather time-consuming

Quality Attribute (QA) A systemic property of a software product, which critically influences its quality

Qualitative approach Approach that tends be general as the population selected is varied and large though there may be low response rate, therefore, does not explain complex issues or interactions; it captures the opinion and attitude of people through interviews, observations, focus groups

Quantitative approach Factual approach that uses scientific or mathematical data to understand a problem, such as analyzing surveys to predict consumer demand

Rapid prototyping, model The activity of creating prototypes of software applications, i.e. incomplete versions of the software being developed. A prototype typically simulates only a few aspects of, and may be completely different from, the final product

Refactoring The process of restructuring existing computer code without changing its external behavior. Refactoring improves nonfunctional attributes of the software

Requirements analysis A software development lifecycle stage, which identifies the desired properties of the future software product

Requirements specification A software development lifecycle stage, which formally describes the properties of the future software product

Resonance A quality of evoking response

Responsiveness of GUI Quality of interface to send a feedback to the user that is informative and well acknowledged

Retirement A software development lifecycle stage, when the product is completely and permanently put out of operation

Return On Investment (ROI) A performance measure used to evaluate the efficiency of an investment or to compare the efficiency of a number of different investments. ROI measures the amount of return on an investment relative to the investment's cost

Robust GUI Interface that is able to cope with errors during execution and cope with erroneous input

Remote Procedure Call (RPC) When a computer program causes a procedure to execute in a different address, which is coded as if it were a normal procedure call, without the programmer explicitly coding the details for the remote interaction

Rule-based approach Approach used in sentiment analysis that classifies text based on the number of positive and negative words using booster words, idioms, emoticons among other classification rules

Scaffolding A variety of instructional techniques used to move students progressively toward stronger understanding and, ultimately, greater independence in the learning process

Scalability The capability of a system, network, or process to handle a growing amount of work, or its potential to be enlarged in order to accommodate that growth

Scope, product scope Features and functions that characterize a product, service or result; product scope defines what the product will look like, how will it work, its features, etc.

Scrum master The facilitator for a product development team that uses scrum, a rugby analogy for a development methodology that allows a team to self-organize and make changes quickly. The scrum master manages the process for how information is exchanged

Scrum methodology, An iterative and incremental agile software development methodology for managing product development

Self-adjustment Adjustment of oneself or itself, as to the environment

Sentiment analysis Analysis that refers to the use of natural language processing, text analysis, computational linguistics, and biometrics to systematically identify, extract, quantify, and study affective states and subjective information

Size, product size Overall size of the software being built or modified

Smart contract A computer protocol intended to digitally facilitate, verify, or enforce the negotiation or performance of a contract

Software-as-a-Service (SaaS) A software licensing and delivery model in which software is licensed on a subscription basis and is centrally hosted

Service Oriented Architecture (SOA) A style of software design where services are provided to the other components by application components, through a communication protocol over a network

"Soft" skills Personal attributes, which indicate a high level of emotional intelligence, such as teamwork, negotiations etc.

Software engineering A set of tasks, methods, tools and technologies used to design and implement complex, replicable and high-quality software systems

Software environment Surroundings for an application; usually includes operating system, database system and development tools

Software process An over-arching process of developing a software product

Software product Merchandise consisting of a computer program that is offered for sale

Spiral, model Systems development lifecycle model, which combines the features of the prototyping model and the waterfall model

Sprint A set period of time during which specific work has to be completed and made ready for review

Stakeholder A person with an interest or concern in something

Structural heterogeneity A property of a dataset, which includes both strong and weak-structured elements

Supervised learning The machine learning task of inferring a function from labeled training data

Support Vector Machine (SVM) A classification algorithm used to separate data into two clusters; using a training dataset with a class label, SVM builds a model that assigns a new instance to one of the clusters

Synchronize and stabilize model, A systems development lifecycle model in which teams work in parallel on individual application modules, frequently synchronizing their code with that of other teams, and regularly debugging, i.e. stabilizing the code

System-of-systems A viewing of multiple, dispersed, independent systems in context as part of a larger, more complex system

Technical constraint A technical limitation or restriction

Tool Computer-aided software, which supports the software development processes and methods; typically used for software development or system maintenance

Unsupervised machine learning The machine learning task of inferring a function to describe hidden structure from "unlabeled" data

Vision, product vision An original idea, clear yet informal, of the fundamental differences and customer values for the future software product as compared to the existing ones, and its benefits after the implementation; desired future state that would be achieved by developing and deploying a product

Virtual Machine (VM) An emulation of a computer system

Votebook system Machine connection system for voting on a private blockchain launched by an election commission; voters have an opportunity to check whether their vote was counted, according to representatives of the project

Waterfall, model Sequential design process, used in software development processes, in which progress is seen as flowing steadily downwards through the phases of conception, initiation, analysis, design, construction, testing, production/implementation and maintenance

Zero Knowledge Proof (ZKP) A method by which one party (the prover) can prove to another party (the verifier) that a given statement is true, without conveying any information apart from the fact that the statement is indeed true

Index

A
Abstract Syntax Tree (AST), 76
Activity, 87
Agile, 17
Agility, 20
Aircraft Communications Addressing and Reporting System (ACARS), 150
Application Programming Interface (API), 36
Architectural heterogeneity, 155
Architecture of Integrated Information Systems (ARIS), 21
Artificial Intelligence, 21

B
Backlog, 155
Backtracking, 58
Bauer, F., 13
Best practice, 123
Bit-torrent, 34
Blockchain, 28
Bloom's taxonomy, 155
Business architecture, 24
Business Process Modeling (BPM), 20–22
Business requirements, 25

C
Common Interface Definition Language (CIDL), 68
Common Object Request Broker Architecture (CORBA), 61
Computer-Aided Design (CAD), 157
Computer-Aided Engineering (CAE), 157
Computer-Aided Manufacturing (CAM), 157
Carnegie Mellon University (CMU), 129
Computer-Aided Software Engineering (CASE), 59
Comfort zone, 161
Common vision, 156
Communication, 19
Compromise zone, 161
Concordia University, 30
Content analysis, 40
Continuous Delivery (CD), 94
Continuous Integration (CI), 94
Corpus-based technique, 38
Crisis management, 17
Crisis zone, 161
Curricula, 124
Customer Relations Management (CRM), 103

D
Data warehouse, 145
Data Base Management System (DBMS), 82, 145
Declarative approach, 57
Design pattern, 123
Design specifications, 25
DevOps, 95
Dijkstra, E., 13
Document Object Model (DOM), 149
Domain Specific Language (DSL), 56
Domain Specific Modeling Language (DSMLs), 76

E
Enterprise Architecture (EA), 24
Enterprise Content Management (ECM), 158
Enterprise Engineering Matrix (EAM), 15, 159
Enterprise Resource Planning (ERP), 159

S. V. Zykov, *Managing Software Crisis: A Smart Way to Enterprise Agility*,
Smart Innovation, Systems and Technologies 92,
https://doi.org/10.1007/978-3-319-77917-1

ESUILang, 79
Ethereum, 34
Extract-Transform-Load (ETL), 145

F
Feedback, 34
Formal methodology, 160
Framework matrices, 40
Functional approach, 58

G
Garbage collector function, 58
Geo-marketing, 111
Google Remote Procedure Call (GRPC), 61, 69
Gromoff, 18
Graphical User Interface (GUI), 73

H
Human-related factor, 18
Hoogervorst, 24

I
Imperative approach, 57
Imperative models, 22
Infrastructure-as-a-Service (IaaS), 94
Innopolis, 124

K
Key performance indicator, 161
K-nearest neighbor, 37
Knowledge management, 124
Knowledge transfer, 124

L
Language, 19
Language Workbench (LW), 76
Language-Oriented Programming (LOP), 76
Lexicon-Based approach, 38
Logical programming, 58

M
Marx, K., 13
Max Entropy, 37
Metacognitive, 148
Metadata, 145
Metadata Object Description Schema (MODS), 146
Meta-knowledge transfer, 162
Metaphor, 14
Microservice, 91
Middleware, 61
Mission-critical software, 55
Monolithic application, 88
Multi-paradigm language, 62

N
Naur, P., 13
Nagel, 18
Naïve Bayes, 37

O
Object-oriented approach, 58
Open Vote Network, 30
Optimization, 23
Oracle, 82

P
Pattern matching, 58
Perlis, A., 13
Pi-calculus (aka π-calculus, process calculus), 114, 116
Platform-as-a-Service (PaaS), 93
Polymorphic function, 58
Prior knowledge, 76
Production Lifecycle Management (PLM), 163
Product vision, 143

Q
Qualitative approach, 39
Quantitative approach, 39
Quantitative data analysis, 40

R
Reflection, 20
Remote Procedure Call (RPC), 60
Resonance, 154
Resonant communication, 132
Rule-based approach, 38

S
SAP, 22
Scaffolding, 165
Scheer, 18
Self-adjustment, 130
Semantic Orientation Approach, 37
Sentiment analysis, 35
Service Oriented Architecture (SOA), 24
Slave singleton, 69
Smart contract, 34
Soft skills, 124
Software architecture, 11
Software-as-a-Service (SaaS), 93

Software engineering, 13
Software process, 130
Software testing, 59
Sprint, 166
Supervised machine learning, 37
System-of-Systems, 166

T
Teaching, 126
Teamwork, 124

U
Unified Modeling Language (UML), 73
University of Maryland, The, 30
Unsupervised machine learning, 37

V
Very Large Database (VLDB), 87
Virtual Machine (VM), 115
Visual Studio (VS). Net, CASE tool, 73
Votebook System, The, 29

Z
Zero Knowledge Proof (ZKP), 31

Printed in the United States
By Bookmasters